SPACE AND TIME IN
SPECIAL RELATIVITY

SPACE AND TIME IN
SPECIAL RELATIVITY

N. David Mermin

Laboratory of Atomic and Solid State Physics
Cornell University

WAVELAND

PRESS, INC.

Prospect Heights, Illinois

For information about this book, write or call:

Waveland Press, Inc.
P.O. Box 400
Prospect Heights, Illinois 60070
(708) 634-0081

ISBN 0-88133-420-0

Printed in the United States of America

7 6 5 4 3 2

My great uncle Ambrose Fudge
said Bernard carelessly.
He looks a thourough ancester said
Ethel kindly.
Well he was said Bernard in a proud tone
he was really the Sinister son of Queen
Victoria.
Not really cried Ethel in excited
tones but what does that mean.
Well I dont quite know said Bernard
Clark it puzzles me very much but
ancesters do turn quear at times.
Perhaps it means god son said Mr
Salteena in an inteligent voice.
Well I dont think so said
Bernard but I mean to find out.
It is very grand anyhow said Ethel.
It is that replied her host geniully.

Daisy Ashford,
"The Young Visiters"
(Doubleday & Co., Inc.,
Garden City, N.Y., 1919.)

PREFACE

The special theory of relativity, alone among the areas of modern physics, can in large part be honestly explained to someone with no formal background in physics and none in mathematics beyond a little algebra and geometry. This is quite remarkable. One can popularize the quantum theory at the price of gross oversimplification and distortion, ending up with a rather uneasy compromise between what the facts dictate and what it is possible to convey in ordinary language. In relativity, on the contrary, a straightforward and rigorous development of the subject can be completely simple.

Nevertheless special relativity is one of the hardest of subjects for a beginner to grasp, for its very simplicity emphasizes the distressing fact that its basic notions are in direct contradiction to certain simple, commonplace notions that almost everyone fully grasps and believes, even though they are wrong. As a result, teaching relativity is rather like conducting psychotherapy. It is not enough simply to state what is going on, for there is an enormous amount of resistance to be broken down.

This book presents an elementary but complete exposition of the relativistic theory of the measurement of intervals in space and time, examining the extraordinary relativistic properties of moving clocks and measuring sticks. At the same time, it attempts to expose the pitfalls and misconceptions the neophyte usually succumbs to.

The book grew out of lectures given to a small class of rather sophisticated high school teachers who met during the spring term of 1964–1965 at Cornell; the course was devoted exclusively to trying to make them believe that special relativity was not inconsistent or paradoxical. I have tried to retain the flavor of the argument that went on in that class, interweaving exposition with remarks about objections or misconceptions that seem likely to arise. This approach seems uniquely pertinent to the teaching of relativity, where misunderstanding is continually generated by commonplace, incorrect notions that are often implicit in the very language we use (a pre-relativistic structure) and are therefore particularly difficult to recognize.

The approach taken here is completely elementary in that no specific acquaintance with physics on the part of the reader is assumed (except in Chap. 18 on energy and momentum) and no mathematical background beyond elementary algebra, plane geometry, and (occasionally) trigonometry. Nevertheless, because of the peculiar nature of the subject, the development of the relativistic theory of space and time is honest and complete, and I have been able to use several parts of the manuscript in teaching a graduate course in classical physics.

I have deliberately dealt hardly at all with the historical or experimental aspects of the subject, which are treated well on this level in many books on modern physics. The historical approach, though fascinating, provides an indirect and frequently confusing route to a clear statement of what the modern theory says. The experimental background and justification of the theory are also intriguing, and without them the entire subject would be a bizarre fairy tale; however they are peripheral to, and even distracting from, the problem here, which is that of reconciling with one's faulty non-relativistic intuition the mere possibility of relativistic effects. Therefore the important experimental results on which the theory rests are simply stated, in sufficient detail to make them clear, but with no attention to the remarkable expenditure of effort and ingenuity behind them. One is in a much better position to appreciate these parts of the subject after one has fully understood the relativistic model of the universe that they imply.

The foundations of the relativistic theory of space and time are developed in Chaps. 1 to 7. All the remaining chapters (except for 18) are primarily elaborations of the basic relativistic effects presented in the first seven. The chapters beyond

7 are largely independent of one another (except that 18 presupposes 13) and need not be read in the order in which they occur, although an increased level of sophistication is assumed as the chapters advance, and detailed explanations become less frequent. Of the later chapters, 13 to 15, 17, and 18 contain further conventional developments of the basic relativistic effects, while Chaps. 8 to 12 and 16 are included for entirely pedagogical purposes—weapons in the battle against disbelief. Finally, Chap. 17, on the Minkowski approach, is entirely self-contained and could, in principle, be read immediately after 1 and 2. I doubt, though, that this would be an easy task.

This book owes its existence in large part to the members of Physics 490, whose active skepticism continually rescued me from that comfortable superficiality and mild dishonesty it is so easy to succumb to. But were it not for the persistent conviction of Lyman G. Parratt that physicists should meet physics teachers, I would never have dreamed of embarking on this enterprise.

N. David Mermin

CONTENTS

Preface   *vii*

Chapter 1   The Principle of Relativity   1

2   The Principle of the Constancy of the Velocity of Light   9

3   How to Approach the Problem   19

4   Length of a Moving Stick (I)   27

5   Moving Clocks (I)   33

Appendix to Chapter 5: Another Way of Proving It   *43*

6   Length of a Moving Stick (II)   47

Appendix to Chapter 6: Another Way of Proving It   *59*

7   Moving Clocks (II)   63

Appendix to Chapter 7: Another Way of Proving It   *69*

8   Objections and Reflections   73

9   Why He Says My Meter Stick Shrinks Although I Know that His Shrinks   79

10   Why He Says My Clocks Go Slowly Although I Know that His Go Slowly   87

11   A Relativistic Tragicomedy   91

12   How Much Is Physics, and How Much
     Simple Arithmetic?   99

13   The Lorentz Transformation   119

14   A Simple Way to Go Faster than Light that
     Does Not Work   129

15   Why Nothing Can Go Faster than Light   135

16   The Clock "Paradox"   141

17   Minkowski Diagrams: The Geometry of
     Space-Time   155

18   Energy and Momentum   201

19   Why?   225

Problems   229

Index   235

SPACE AND TIME IN SPECIAL RELATIVITY

1
THE PRINCIPLE
OF RELATIVITY

The special theory of relativity rests on two experimental facts. One is a familiar part of the everyday experience of travelers, while the other cannot be directly perceived and was established only by very sophisticated and precise measurements. Reflecting this, the first fact, known as the principle of relativity, was recognized by Galileo in the early seventeenth century, but the second, known as the principle of the constancy of the velocity of light, was not suspected until the late nineteenth century and was established convincingly by experiments only in the early twentieth century.

These facts, or principles—a title they fully deserve because of their apparent universality—differ not only in their degree of remoteness from common experience, but also in the extent to which they excite or surprise. The principle of relativity, though not believed by Aristotle and shocking to many when it first gained currency, now belongs to the body of general common knowledge and arouses disbelief or a sense of paradox in very few. On the other hand, experiments leading to the enunciation of the principle of the constancy of the velocity of light, when first performed less than a century ago, were considered at best to be surprising and puzzling and at worst, completely unintelligible.

The mysterious aspects of the second principle were resolved in 1905 by Einstein, who showed that they were due to naïvely assuming some apparently obvious properties of space and time which were in fact grossly incorrect when applied to anything moving at very great speeds. By accepting both principles, he was able to deduce a simple, novel, and consistent picture of the world. This new view hardly differs from older notions when extremely rapidly moving things are not considered, but it leads to unexpected and remarkable predictions whenever enormous speeds are involved. Many of these predictions have been confirmed, and none have yet been contradicted.

Before seeing how Einstein's theory, the special theory of relativity, illuminates the principles, we must state and examine them. The principle of relativity, the more subtle but less astonishing of the two, is considered in this chapter, and the constancy of the velocity of light, simple but at first glance absurd, is taken up in Chap. 2.

Galileo stated the principle of relativity in the "Dialogue on the Great World Systems." It occurs in his refutation of an argument of Aristotle's that the Earth stands still. According to Aristotle, the Earth cannot be moving because a ball thrown straight up eventually falls to Earth at the point it was thrown from. If the Earth were moving, it would move while the ball was aloft, so that as it landed, some new part of the Earth would be beneath the ball.

Galileo pointed out that this reasoning is wrong because if the Earth were moving to one side, the ball, originally on the Earth, would be moving in the same way. When thrown into the air it would, in addition, have a new vertical motion, but at the same time it would continue in its original sideways motion. Thus although the Earth would indeed move to the side as the ball went up and down, the ball would move to the side by the same amount and come down on the same part of the Earth from which it was thrown.

The conclusion is that nothing can be learned from Aristotle's experiment. Whether the Earth moves slowly, rapidly, or not at all, the ball will still land in the same place it was vertically tossed from. This observation is an instance of a general principle: One cannot tell by any experiment whether one is at rest or moving uniformly.

Galileo illustrated the principle by contemplating a laboratory shut up in the hold of a ship, in which a variety of biological and physical experiments could be performed, none of which would give any indication whatsoever of the velocity of the ship as long as the ship's motion was perfectly smooth and uniform. Goldfish would continue to move in a random manner inside their bowl, drops from leaky faucets would land in the usual spots, and flies would continue to bother one from random directions, all regardless of the motion of the vessel.

One encounters the same principle in moving trains or airplanes, provided the motion is truly uniform and there is no jostling or bouncing. Flying smoothly at 500 miles per hour, one observes that coffee poured from a pot falls quietly into the cup below and that dropped objects fall directly down to the floor. One is aware of motion only when looking out the window at clouds or at the ground. Nothing within the moving plane behaves differently from the way it would were the plane at rest or, for that matter, moving with any speed other than the one it has. Alternatively, nothing done within a plane that is either at rest or moving uniformly gives any clue as to whether the plane is at rest or moving uniformly or as to the particular speed with which it moves.

This is certainly not obvious, nor the kind of thing one could deduce by sheer logic. It is an experimental fact, which has been repeatedly confirmed. One can easily imagine that it might not be true. One could suppose, for instance, that a chemist might concoct a liquid that boiled at 150 degrees in his laboratory but that when placed on a train moving past his laboratory boiled at 145 degrees when the train moved at 20 miles per hour, at 140 degrees at 40 miles per hour, at 135 degrees at 60 miles per hour, etc. There is nothing inconceivable in this. However no such substance has ever been found. Provided the pressure in the train is the same as the pressure in the laboratory, if a liquid boils at 150 degrees in the laboratory, it will also boil at 150 degrees in the train, however fast the train is going. No one has ever produced a substance or device that behaves differently in a laboratory at rest from the way it would behave in a uniformly moving laboratory: Things happen in a uniformly moving laboratory in exactly the same way that they happen in a laboratory at rest. The principle of relativity is a statement of this inability to distinguish between states of rest or uniform motion. It can be stated, as before:

One cannot tell, by any experiment, whether one is at rest or moving uniformly.

An equivalent way of putting it is this:

If two experiments are performed under identical conditions except that one is done in a laboratory at rest and the other in a uniformly moving laboratory, the two experiments will lead to exactly the same conclusions.

If we refer to the accumulated results of all experiments as the laws of nature, a third way of stating the principle of relativity is:

The laws of nature are the same in a laboratory at rest as they are in any uniformly moving laboratory.

These are three equivalent ways of expressing the impossibility of distinguishing between states of rest and states of uniform motion; they all say the same thing, but it usually helps to have several equivalent versions of a basic principle.

One should realize the importance of the word "uniformly," which has quietly appeared in all three formulations. If a laboratory, train, or boat is moving, but not uniformly, experiments can detect this. If a ship moves steadily on a perfectly calm sea, shut up in a cabin one is completely unaware of the motion. If, however, waves bounce the ship up and down as it moves, one is immediately aware of the nonuniform motion. Similarly, things behave as they do at rest in a train that progresses smoothly at constant speed in a straight line, but if the train veers off to the left, objects hanging from the ceiling will swing to the right, tea will slosh in cups, and standing passengers will have to brace themselves to preserve their balance, none of which phenomena could happen in a stationary or uniformly moving train. Only states of motion in a straight line with constant speed are indistinguishable from states of rest.

Another point to notice is that since a state of uniform motion is indistinguishable from a state of rest, there is no experimental basis for calling the state of rest a state of rest and the state of motion a state of motion. Suppose laboratory A is at rest and laboratory B moves uniformly. Since all experiments done in either give the same results, if a man in B asserts that on the contrary he is at rest and laboratory A moves uniformly in the opposite direction, there is no way to show him that he is wrong. Indeed, he is not wrong. It is a matter of taste or convention. Everything we can say about experiments done in the two laboratories will be the same whether we choose to regard A as at rest and B as moving uniformly with a certain speed, B as at rest and A as moving uniformly in the opposite direction with the same speed, A as moving one way and B as moving the other with equal and opposite speeds, or any other way that is consistent with the fact that the distance between A and B grows at a constant rate. Arguing about which of the two is really at rest is rather like arguing over whether the South Pole is 8,000 miles directly under the North Pole or, on the contrary, the North Pole is 8,000 miles directly under the South Pole. Either way of putting it is as good as the other, and one is free to choose the way that suits one's convenience, though one should always be aware of the arbitrariness of the choice.

This degree of freedom in deciding who is at rest is so important that it is worth stating the principle of relativity in yet another way that emphasizes this aspect:

Anybody moving uniformly with respect to somebody at rest is entitled to consider himself to be at rest and the other person to be moving uniformly.

This particular formulation (which is equivalent to the other three) accounts for the name "relativity." It denies that there is any absolute meaning to the notion of being at rest and asserts that when we say that something is at rest, we must specify relative to which of all the innumerable, equally good, uniformly moving laboratories the thing is at rest. If we wish to be precise, we should not categorically assert that Mr. Wilkins is at rest; rather, we should say something like "Mr. Wilkins is either at rest or moving uniformly, depending on which of all the equally good laboratories you choose to measure his speed from." This makes for rather long-winded sentences, and so it is useful to have a single phrase that describes a state of rest or uniform motion without committing us to whether we wish to consider the state to be one of rest or one of motion. A laboratory that is in either a state of rest or uniform motion is called an *inertial laboratory* or an *inertial system* or an *inertial frame of reference*. Instead of saying "Mr. Wilkins is at rest," we shall say "Mr. Wilkins is in an inertial system" or "Mr. Wilkins is in an inertial frame," which sounds a bit clumsier but reminds us that the distinction between rest and uniform motion is arbitrary.

The word "inertial" gives us a fifth and rather concise way of stating the principle of relativity:

A system moving uniformly with respect to an inertial system is also an inertial system.

For "system" one can also read "frame of reference" or "laboratory" or—if one is thinking of the person performing experiments—"observer." If Mr. Wilkins looks around him in an inertial system, he is an inertial observer.

Before concluding this chapter on Galileo's principle of relativity, I should mention why the peculiar word "inertial" was chosen to describe something one is entitled to regard as either moving uniformly or at rest. The word reflects the kind of issues that arise if one asks questions like "Precisely what is meant by a state of rest?" I have only said up to now that if one person is in a state of rest, any person moving uniformly with respect to him is also entitled to regard himself as in a state of rest. But I did assume that we could all identify at least one state of rest. One can spend considerable time worrying about this question, and a thorough, honest analysis of it would lead one into many profound issues. However it is a mistake to become sidetracked here, when first approaching relativity. One has a fairly clear intuitive notion of when something can be considered at rest or moving uniformly, and special relativity can be perfectly understood in terms of this intuitive idea. The question of how to give a precise definition of a state of rest (or an inertial frame) is a side issue, and we shall not be distracted by it. The following elaboration of the intuitive notion is completely adequate for our purposes:

A state of rest is one in which Newton's first law of motion holds. This law (known as the law of inertia, whence the name "inertial frame") states that a body which is not acted upon by any forces either does not move or moves with constant speed in a fixed direction. One is therefore in an inertial frame if, when no forces act on a body, the body either stands still or moves uniformly along a straight path.

That will be enough for us. Those who worry over how one can be sure that no forces act are grappling with a subtle and difficult question, but it has virtually no bearing on the problem of understanding special relativity. For that purpose a simple, unsophisticated understanding of the principle of relativity is sufficient:

One *can* tell by experiments if one is in a laboratory that moves nonuniformly (for example, jerkily along a curved path or with varying speed in the same direction), but if the experiments indicate that the laboratory is moving uniformly, i.e., with constant speed along a straight line, there is no further experiment that will tell us what the value of this constant speed is, i.e., whether it is zero, 1 inch per second, or 50,000 miles per second.

2

THE PRINCIPLE OF THE
CONSTANCY OF THE
VELOCITY OF LIGHT

The birth of special relativity was stimulated by experiments on the nature of light. Indeed, in its early days it looked as if the emerging theory of relativity might be just a part of the very extensive theory of light and electromagnetism. It was Einstein who realized that in fact relativity is far more general, embracing the behavior of all natural phenomena.

The reason for this close historical association between relativity and the properties of light is that relativistic effects are only noticeable in the behavior of things moving with enormous speeds. In the nineteenth century, the only thing available that moved fast enough and had properties that could be examined in detail was light. Today relativistic effects can be observed in the behavior of cosmic rays and of electrons, protons, and other particles pushed to immense speeds in high-energy accelerators. At the turn of the century these phenomena were unknown, and it was only through the behavior of light that the first clue was received that something was very wrong with our ideas about space and time. Not the least part of Einstein's contribution was his understanding that certain deeply puzzling results arising from experiments on light had an explanation that had to do not so much with the nature of light as with the nature of space and time.

The only aspect of light we shall have to discuss here is one very strange property of its experimentally measured speed. Until informed to the contrary, one tends to think of light as moving with infinite speed, i.e., as getting from here to there in no time at all. This, as just about everybody now knows, is not true. (Galileo appears to have been the first person to have tried to establish this experimentally. He failed—the speed was too great for his method.) When we turn on a lamp, a tiny interval of time elapses between the moment the light leaves the bulb and the moment it reaches our eyes. This is because light does not move with infinite speed, but with a speed that is finite, though enormous.

Because the speed is so large, a delicate and sophisticated experiment is needed to measure it, but it can be measured. Light moves through empty space at a speed that is very close to 300,000 kilometers per second, or about 186,000 miles per second. (If there is matter present, light will go at a somewhat different speed. In air it goes a little slower than 300,000 kilometers per second and in water or glass, considerably slower. By "the speed of light" we shall mean the speed at which light travels in a vacuum.) Today this speed has been measured with considerable precision and is known to be about 2.9979 hundred thousand kilometers per second. This particular velocity—the speed of light in empty space—is of very basic importance and a special symbol is reserved for it. The letter c stands for the speed at which light moves through empty space: $c = 2.9979$ hundred thousand kilometers per second.

The remarkable, indeed, the shocking, thing about the speed of light is this: When one says that light moves with a speed c through empty space, it appears that one has not said enough to specify how fast a particular beam of light will go past a particular observer. For example, if I remarked that a fellow passenger was walking toward the front of our plane at 4 miles per hour, I would mean, of course, at 4 miles per hour with respect to me, an observer in the moving airplane. If the plane were moving at 400 miles per hour, the man would be moving at 404 miles per hour past an observer on the ground and at 804 miles per hour past an observer in a plane going 400 miles per hour in the opposite direction. This leads one to suspect that when one says the speed of light is c, one must also say with respect to what kind of observer it moves with velocity c. One is tempted to say that it must be an observer who is at rest, but the principle of relativity tells us that any observer moving uniformly with respect to an observer at rest can also be regarded as being at rest. The extraordinary thing is, however, that this does not matter. There is overwhelming experimental evidence that *the speed of light has the same value c with respect to any inertial observer.*

This fact is so surprising and hard to accept that after hearing it for the first time, most people either miss the point or, if they have understood it, think that they must have misunderstood. Let me say it again in a somewhat more concrete way. Suppose I am in an inertial frame and consider myself, as I am entitled to do, to be at rest. A flash of light passes by me. If I measure how fast it goes past with my very accurate and expensive modern equipment, I shall find that its speed is c. Suppose that you are moving past me in the direction of the light at 15 kilometers per second in your space capsule and that you also measure the speed of the same flash of light with your equipment. What answer will you get?

Until the late nineteenth century everybody would have agreed in predicting that your answer would be c minus 15 kilometers per second. We now know that this is wrong. You will also find that the same light goes past you with exactly the same speed c as I found that it went past me. *Any* inertial observer will get the same answer: c. Someone who zooms past me at the enormous speed of nine-tenths of the speed of light in the direction of the flash will find, if he measures the speed of the flash, that it does not go past him at $1/10\ c$, but at the same old speed: c itself; if he goes past me at $9/10\ c$ in the opposite direction to that of the flash, he will find when he measures the speed of the flash that it is not $19/10\ c$, but still c. The result does not depend on the speed of the observer; all inertial observers, no matter how fast they are going* or in

> * This is the first in a series of notes containing further elaborations, explanations, and clarifications or pointing out subtleties or ambiguities that may not be immediately evident. If I included all such remarks in the text it would make for a rather choppy and meandering analysis. These notes should be regarded as an additional commentary on the text, dealing with questions or sources of confusion that might arise or pointing out distracting questions that should have been mentioned, but which were glossed over or even ignored for the sake of even exposition. If the notes prove to be too distracting, I would suggest that you read the chapter once ignoring them, and then read it again with the notes. The purpose of this particular note is to mention that those who are baffled at this point by the thought of what an observer moving in the direction of the light with a speed *greater than c* would see will perhaps be reassured to learn that they have good cause to be baffled. The answer to their question is (Chap. 15) that no observer can move so fast.

what directions, measuring the speed of the same flash of light will all come up with the same answer: c. The experimental evidence both direct and indirect for this extraordinary fact is now so extensive that it can be regarded as one of the basic laws of nature. It is known as the principle of the constancy of the velocity of light.

The fact of the constancy of the velocity of light did not, of course, spring self-evident from the first laboratory to perform the appropriate experiments. People thought at first that light moved with a speed c with respect to a definite something which they called the ether.* If this were so,

> * The ether was also invoked to account for the wavelike properties of light. Light was thought to be an oscillation of the ether.

pre-relativistic reasoning led them to expect that if the Earth moved through the ether with a speed v, the speed of light moving past the Earth in the same direction as the Earth's motion through the ether would be $c - v$ when measured from the Earth, the speed of light moving past the Earth in the opposite direction from the Earth's motion through the ether would be $c + v$ when measured from the Earth, and in general the speed of light with respect to the Earth in any arbitrary direction would depend on the angle between that direction and the direction of the Earth's motion through the ether.

In essence, the famous Michelson-Morley experiment* was

* A concise description of the Michelson-Morley experiment, together with a summary of the various non-relativistic attempts to account for its outcome, can be found in W. Panofsky and M. Phillips, "Classical Electricity and Magnetism," chap. 14, Addison-Wesley Publishing Company, Reading, Mass.

an attempt to measure this directional dependence of the speed of light with respect to the Earth and thus to determine the speed of the Earth with respect to the ether. The result of their experiment was that the speed of light with respect to the Earth has the same value c whatever the direction of motion of the light.

One might try to explain this by saying that the speed of the Earth with respect to the ether must be zero. Aside from the fact that this would be a rather strange coincidence, this explanation will not do. The Earth moves in its orbit around the Sun at about 30 kilometers per second. If the velocity of the Earth with respect to the ether happened to be zero at one time of year, then 6 months later when the Earth was moving at the same speed but in the opposite direction, its speed with respect to the ether would have to be 60 kilometers per second. In general, because of the Earth's motion around the Sun, whatever the speed of the Earth with respect to the ether might be, this speed should vary through a range of speeds differing by up to 60 kilometers per second, throughout the course of a year. However experiments have shown that the speed of light with respect to the Earth is independent of direction, whatever the time of year.

Thus if the ether does exist, it must be managing in a most mysterious way to escape our efforts to detect it. As Einstein showed, the way out of this dilemma is to deny the existence of the ether and face courageously the fact that light moves with a speed c with respect to any inertial observer whatsoever, regardless of the velocity of that observer with respect to any other observer (and, as a special case of this, regardless of the velocity of the source emitting the light).

The failure of all other, less revolutionary attempts to account for this experimental result is described in a number of places, to which I refer you* in order that you may be con-

> * Chapter 14 of Panofsky and Phillips contains a list of references describing such theories.

vinced that the shocking principle of the constancy of the velocity of light was not propounded lightly, but only after all other arguments that anybody could devise to explain the experiments had been shown to be inadequate. To do justice to the more conventional explanations, I should devote a chapter to them, but since their only significance today is that they have all been shown to be wrong, the only point of such a chapter would be to make you more willing to consider a proposal as drastic as the constancy of the velocity of light. Since the aim of this book is to show directly that the constancy of c is not as shocking as it first appears, it seems more sensible to go about this at once. I shall ask you, however difficult it may at first seem, to accept the principle of the constancy of the velocity of light as another fact of nature, just as I have asked you to accept the (apparently less exotic) principle of relativity. Given these two experimental facts, we shall see how, when taken together, they lead to a simple, consistent picture of the world, with many new, beautiful, and surprising aspects.

To avoid any possible misunderstanding of this principle of the constancy of the velocity of light that experiments have forced us to accept, let me state it again, as baldly as I can. One should realize, to begin with, that the principle does not find its explanation in any peculiar, complicated, or subtle theory of what light in fact is. It is true that light is, under the most careful analysis, very peculiar, complicated, and subtle, but this has little bearing on the principle. We can understand the principle by thinking of a beam of light as simply a collection of many little particles, called photons,* moving along

* The fact that light can be regarded as either a wave or a beam of particles is an aspect of the quantum theory.

at the speed c. It is an experimental fact that any photon, i.e., any little particle of light, moves with a speed c with respect to any inertial observer at all. This is quite unlike the behavior of any of the things one is used to thinking about. If a thrown ball moves at 50 miles per hour away from the person who threw it and if I run after the ball at 10 miles per hour, I reduce its speed with respect to me to 40 miles per hour. If I drive after it in a car, I can reduce its speed with respect to me still more. If I drive after it at 50 miles per hour, it has no speed at all with respect to me; i.e., it does not get any farther away from me. If I drive faster than 50 miles per hour, I can actually catch up with and overtake the ball.

Photons are quite different. No matter how fast I drive I cannot catch up with a photon. I cannot even begin to reduce the rate at which it recedes from me. However fast I chase after it, I shall find that the distance between me and the photon continues to grow at exactly the same rate as before: 2.9979 hundred thousand kilometers per second.

Notice that however baffling this property of light may be, it has one clearly pleasing feature. If the speed at which a photon moved through *empty* space were not c in all inertial frames, but only in a particular one, we should have to abandon the principle of relativity, for we could then distinguish between different inertial laboratories by experiments performed entirely within those laboratories. One would only have to make some photons (by turning on a light) and see how fast they moved through the laboratory. The answer would be c only in one inertial frame, and in general people would get different answers depending on how fast they were going with respect to this particular frame. The fact that the answer is c in all inertial laboratories shows that the behavior of light is consistent with the principle of relativity, i.e., with the impossibility of getting different answers from identical experiments performed in laboratories moving with respect to one another.

In the following chapters we shall be talking about light quite a lot, but this should not mislead you into thinking that special relativity is primarily a theory of light. Light keeps appearing in modern discussions of relativity only because it is the most familiar thing that moves with the speed c. It is not, however, the only thing; neutrinos, for instance, also move with the speed c in vacuum, as far as anybody knows today. There is also a principle of the constancy of the velocity of neutrinos. We could just as well phrase all the following discussions in terms of neutrinos and never mention light again. We use light instead because it is more familiar and easier to imagine (and because some of the devices and experiments we shall examine could, in principle, be done with light much more easily than with neutrinos).

Thus although light appears in a discussion of special relativity as the best-known example of something that moves with the speed c, the interesting thing about the speed c is *not* that it is the speed of light. On the contrary, the only interesting thing about light for our purposes here is that its speed is c. The speed c is the thing of fundamental significance. We stated the principle of the constancy of the velocity of light as follows:

The speed of light has the same value c with respect to any inertial observer.

However it could equally well be stated without any reference to light:

Anything that moves past a given inertial observer with the speed c moves past any other inertial observer with the same speed c.

This appears to be a much more general assertion than the principle of the constancy of the velocity of light, but in fact it follows directly from that principle. Suppose some other kind of particle, a blip, moved with the speed c in some particular inertial frame. Consider an experiment that asks the question: Which particle wins a race in empty space, the photon or the blip? In an inertial frame in which the speed of blips is c, the outcome of the experiment is that the race is always a tie, since the speed of light is c in any inertial frame. Now the principle of relativity tells us that if the same experiment is done in two inertial laboratories, it must lead to the same conclusion regardless of the relative velocities of the two laboratories. Hence the outcome of such an experiment in any other inertial laboratory must also be that the race is a tie. Since the speed of light is c in any inertial laboratory, the race can be a tie in every inertial laboratory only if the speed of blips is also c in every inertial laboratory.

3

HOW TO APPROACH
THE PROBLEM

Our task is to make sense of the experimental fact that if any two inertial observers measure the speed of the same flash of light, they will both find that it moves with the same speed c, even though the two observers may themselves be moving with respect to each other. The only other experimental fact we have at our disposal is the principle of relativity—that any experiment performed in a laboratory uniformly moving or at rest must give the same result when it is performed in a laboratory moving uniformly past the first one.

Remarkably enough, by using these two facts alone,* we

* This is a slight overstatement; we shall need a few other minor principles which I shall try to point out as they arise; they are all, however, of so obvious a nature that if I did not call your attention to them, you might not even notice them. Nevertheless, since we shall see many "obvious" assumptions turn out to be wrong, we should try to keep all our assumptions, "obvious" or not, out in front of us.

can solve the problem. By applying the principle of relativity with some ingenuity, we can arrive at a detailed picture of what the world must be like in order for the principle of the constancy of the velocity of light to be possible. The picture, which follows from these two principles enables us not only to understand why c is the same for all inertial observers, but also to predict other unexpected and remarkable properties of the natural world.

To see where to begin, let us sharpen our intuitive disbelief in the principle of the constancy of the velocity of light by trying to construct a "proof" that the principle is incorrect.

Suppose a flash of light passes an inertial observer, whom we shall call A, with a speed A measures to be c. Suppose a second observer B moves past A in the direction of the light at a speed which, to make things simple, we assume A measures to be half the speed of light, $c/2$. Our everyday intuition (and pre-relativistic physicists) would immediately conclude that if B measured the speed at which the light receded from him, he would find the answer $c/2$. If pressed, we could justify this conclusion as follows:

Suppose at the beginning A, B, and the light flash are all in the same place. Subsequently A finds that the light moves away from him at a speed c, while B moves away in the same direction at half that speed. Consequently after any given length of time has elapsed, the light is twice as far from A as B is. Another way of putting this is to say that after any given length of time t has elapsed, the distance d from A to B is the same as the distance from B to the light. But the speed of B past A is just the distance d from A to B divided by the time t it took B to go that far. In the same way the speed of the light past B is just the distance d from the light to B times the time t it took the light to get that far from B. Thus the speed of B past A, which we are told is $c/2$, must be the same quantity d/t as the speed of the light past B. Hence the speed of the light past B must be $c/2$, which is half the speed of the light past A.

This tedious and cumbersome expansion of one's intuitive conviction that if one chases after light with half its speed, the light escapes only half as fast, is wrong. It was presented in all its grim detail to lay bare the area in which the mistake must lie. For the experimental fact is that if B makes the measurement he will find that the light moves past him with the same speed c as it moves past A.

The mistake in our analysis is a subtle one. It must be emphasized, to begin with, that A does not merely find in some ethereal way that B and the light have their respective speeds; he must reach this conclusion by actual measurements using material devices that measure time intervals (clocks) and material devices that measure distances (meter sticks). Similarly, the simplest way for B to measure the necessary times and distances is by using his own clocks and meter sticks, which move with him. In the foregoing analysis we have made some assumptions about the clocks and meter sticks used by B that probably struck you as so obviously correct that you failed to notice them. We assumed that when A's clocks tell him a time t has elapsed between two events, then B's clocks tell B that the same time t has elapsed between the same two events. We assumed that when A's measuring sticks tell him that two things are a distance d apart, then B will draw the same conclusion using his own measuring sticks. We assumed that when A finds from his clocks that two things happen at the same time (in this case, the light passing a certain point while B passes another point half as far from A), then B's clocks will also lead B to conclude that the same two events happened at the same time.

Only if all these assumptions are correct are we entitled to conclude from A's measurements what the outcome of B's measurements will be. If any of them is wrong, we cannot so easily conclude that the light will pass B with a speed $c/2$.

We shall see that, in fact, *all* these assumptions are wrong. In the real world B's clocks and meter sticks (or whatever other devices he uses to measure the speed of the light past himself) will, because they move with B, behave differently from A's equipment. This difference is just enough to cause B to conclude that the light is going past him with the speed c, rather than $c/2$.*

* It may occur to you to worry at this point that if B's clocks and meter sticks behave differently from A's, the results of the same experiments performed by both B and A would be different, thus contradicting the principle of relativity. We shall see that this is not so. On the contrary, given the principle of the constancy of the velocity of light, it turns out that the principle of relativity *requires* moving clocks and meter sticks to behave differently from stationary ones.

It is best not to worry at once about what B's clocks and meter sticks are actually doing and how they can be doing it or whether this is something profound or merely a desperate attempt to escape the dilemma. We shall be in a much better position to consider these matters after the next four chapters, in which all the peculiar aspects of moving meter sticks and clocks will be examined in a specific and quantitative way. At the moment the subject was raised only to explain the direction we must take: If the principle of the constancy of the velocity of light is right, something peculiar must be happening to moving meter sticks and clocks. Our immediate task is to find out what.

This will keep us occupied in Chaps. 4 to 7, but a few more preliminary remarks are necessary. The first concerns a linguistic matter. Language can be enormously confusing when it is used to discuss relativity, because words and even grammar often introduce physical assumptions into what we say with such subtlety that we fail to realize that the assumptions are present. This can be quite useful when the assumptions are correct, but it can be disastrous when, as is frequently the case, the assumptions are wrong. For example, the sentence "New York is about 1,000 miles away" has an illusory absolute quality about it which we have all learned to reject. We know that the sentence is true if spoken by someone in Iowa, and false if spoken by someone in Australia. What is less obvious, but just as true, is that a sentence like "Five minutes after the game began a touchdown was scored" can also be true or false depending on the speaker (but in this case to determine the truth of the sentence we must know not where the speaker is,* but how fast he is going past the stadium).

* Relativistically this is an ambiguous statement, even when all speakers are referring to the same game.

If one forgets things like this, language frequently can seduce one into taking statements that are true for one speaker (or observer) and assuming incorrectly that they will also be true for another.

To avoid confusion, we shall frequently have to expand statements to include a specification of the observers for whom the statement is correct, even though to non-relativistic ears this sounds like an unnecessary piece of information. I shall not do this all the time because it leads to tedious and awkward constructions which, after one has learned to be careful, are frequently unnecessary. For example, when I say "Consider a stationary meter stick," what I really mean is "Consider a uniformly moving meter stick from the point of view of an inertial frame in which it is at rest." If there is a possibility of confusion, I shall try to use the more precise, if clumsier, form.

We shall run into this problem in cases more subtle than the example just given, frequently finding that sentences that seem to have precise meanings are actually quite ambiguous until the frame of reference (or laboratory) considered to be at rest is specified. If you think that this point is a pedantic quibble, consider the fact that in the light of relativity theory, sentences containing phrases like "a distance d apart," "happens at time t," "before," "after," and "at the same time" are all as ambiguous as those containing the phrases "at rest," "moving with speed v," and "5 miles from here," until the observer making such statements is specified. (However the phrase "moving with speed c" does not produce any such ambiguity. That is another, rather amusing way of looking at the principle of the constancy of the velocity of light. When I say "moving with speed c," it is not necessary to specify with respect to which inertial observer.)

One final point should be made before we begin. We are now going to see what we can conclude about the behavior of moving measuring sticks and clocks, while trying to make as few unjustifiable assumptions as possible, i.e., while trying to assume very little more than the principle of relativity and the principle of the constancy of the velocity of light. We shall therefore not assume that moving meter sticks have the same length as stationary meter sticks* nor that moving clocks be-

> * This is a good example of the kind of abbreviation I had in mind. What I really mean is that we shall not assume that a given observer finds that a meter stick moving past him has the same length as a meter stick that is not moving with respect to him.

have in the same way as stationary clocks do. We shall try to figure out how moving clocks and meter sticks actually behave, by using the principle of the constancy of the velocity of light and by appealing continually to the principle of relativity to enable us to consider the behavior of these moving sticks and clocks both from the point of view of some stationary observer and from the point of view of an observer moving along with the sticks and clocks.

We shall be very cautious in assuming that any of the measurements or observations of two such observers agree. We shall, however, assume that when one inertial observer says that two events happen in the same place at the same time (i.e., right on top of each other), any other inertial observer will agree that they happen in the same place at the same time. The second observer may indeed think that the two events happen at a different place and at a different time from the place and time the first observer said they occurred at, but he will say that the two events each occur at the *same* place and time (whatever that place and time may be) if the first observer said this.

This is a "principle of the invariance of coincidences." When one observer says two events coincide in space *and* time, so will all other observers. Note that the word "and" is very important. If one observer finds that two events happen at the same place but at different times (or at the same time but at different places), we shall see that there are observers moving with respect to the first who find that the two events happen at both different places and different times. It is only when two events happen both at the same place and at the same time that all observers agree on their coincidence.

The reason we take this extra principle to be true is that when two things arrive at the same place at the same time, they can have a direct and immediate effect on each other. (Consider, for example, two cars, one moving north and one moving south, arriving at the same place at the same time.) As a result of such conjunctions noticeable things happen. It would be absurd if there were one observer who said the two things were never at the same place at the same time and another observer who said there were definite modifications in the things (e.g., dents in the cars) that could only have arisen from their having been in the same place at the same time.

So much for preliminary remarks. We must now find out how two observers moving with velocity v with respect to each other can each find that the same beam of light is moving at 2.9979 hundred thousand kilometers per second with respect to themselves without one of them being mistaken. Since velocity measurements involve measuring a length (the length traversed) and a time (the time it takes to traverse that length) and since such measurements are made with clocks and meter sticks, we must consider how clocks and meter sticks behave when standing still and when in motion.

We shall consider several simple experiments which can be done in principle with a large stock of clocks and meter sticks. We first check that all the meter sticks are the same by lining them up on top of each other. We can also put all the clocks right next to each other and, after observing them for a long time, check that they all tick at the same rate. Having done this, we are ready to set some of the clocks and meter sticks into motion and figure out ways of measuring their rates and lengths in terms of the stationary clocks and meter sticks.

4

LENGTH OF A MOVING STICK (I)

This will be a short, unexciting, but very important chapter. I want to convince you that if a meter stick moves with uniform velocity along a line perpendicular to its length (like the vertical mast of a ship moving horizontally through a smooth sea), the length of the meter stick will be the same as the length of a meter stick at rest.* This may strike you as being

---

* In accordance with our convention about the use of words like "moving" or "stationary," what the last sentence in the text should really say is, "If a meter stick moves perpendicular to its length with uniform velocity with respect to some particular inertial observer and if that observer measures the length of the meter stick, he will find that it has the same length as a meter stick that is not moving with respect to himself." In stating it in the simple way I did in the text, what I was really doing was to describe the result from the point of view of that particular inertial observer, without explicitly pointing out that I was doing so. Such implicit observers will be lurking behind most of my remarks and you

> should watch for them, since I cannot continually digress like
> this to call them to your attention.

so obvious that it requires no justification. To convince you
that, on the contrary, it is a fact that should be viewed with
suspicion until it has been well established, let me point out
that it is only true when the meter stick moves along a line
exactly perpendicular to the stick. If the meter stick is not
perpendicular to its direction of motion (i.e., if the mast of the
ship does not make a right angle with the deck), the result is
no longer correct. We shall see this in some detail in Chap. 6
which discusses the case of a meter stick moving along a line
parallel to itself (i.e., when the mast of the ship is lying
horizontally on the deck and pointing from stern to bow).

Once we have established the result of this chapter, it will
follow immediately that a moving clock runs slowly compared
with a stationary one* (Chap. 5). Thus when one accepts the

> * Translation into more precise language: Any inertial ob-
> server finds that a clock moving past him with uniform veloc-
> ity runs slowly compared with the rate of clocks that are not
> moving with respect to him.

fact that a meter stick moving perpendicular to its length has
the same length as a stationary meter stick, one is committing
oneself to considerably more than one thinks. Therefore we
had better be quite sure that this first result is right.

We want to measure the length of the moving meter stick
by comparing it with a stationary stick. This, as we shall see,
is rather tricky when the stick moves parallel to its length, but
it can be done in a straightforward way when the stick moves
perpendicular to its length. We just take a second meter stick
and set things up as in Fig. 4.1, with the two sticks parallel
to each other, and a line parallel to the direction of motion
passing through the centers of both.

Let us call the stationary stick A and the moving stick B.
There is a single instant of time when the moving stick crosses
the stationary one, at which moment they are on top of each
other.* At this moment we note the points on A that are in

> * This remark, which may appear obvious, requires some

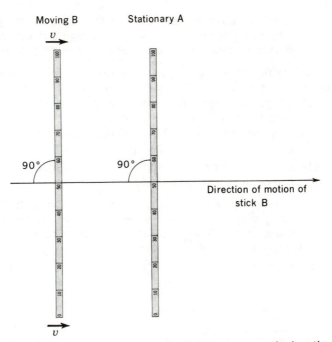

Fig. 4.1   The stationary meter stick A measures the length of stick B, which moves with speed v in the direction of the arrows.

justification since it is an assertion about the simultaneous occurrence of events that are separated in space. That is, we are saying that any point of B passes by A *at the same time* as any other point of B passes A. We shall see that the remark is only correct for observers moving along the line perpendicular to the two sticks (which is all that is necessary for the present argument). For these observers it must be true on grounds of symmetry. If, for example, the 100-centimeter mark of stick B crossed that of stick A *after* the zero-centimeter marks had crossed, there would be no general way of accounting for this that would not also predict that the 100-centimeter marks crossed *first*, since the whole configuration (Fig. 4.1) is symmetric about the line of motion. Such a paradox can only be avoided if both crossings occur at the same time. This argument fails when the sticks are not perpendicular to the line of motion, for one can then distinguish between the two ends, of the stick according to whether they point in the forward or backward direction. (If this note mystifies you, reconsider it after reading Chap. 7.)

contact with the zero-centimeter and 100-centimeter points of B. If these points are the zero-centimeter and 100-centimeter points of A, the length of the moving stick is evidently the same as the length of the stationary stick.

Now suppose, on the contrary, that the moving stick shrank (Fig. 4.2) so that, for example, its 100-centimeter and zero-centimeter marks coincided with the 95-centimeter and 5-centimeter marks on A.* We can invoke the principle of relativity

* Note that here we are assuming that if the top half of the moving stick shrinks, the bottom half shrinks by the same amount. If this were not so, things would be happening differently in one spatial direction from the way they happen in another (in this case opposite) direction. But we shall assume

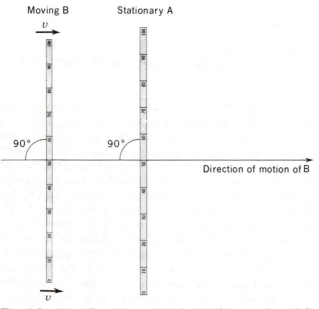

Fig. 4.2   What Fig. 4.1 would look like if the moving stick shrank.

that things behave the same way regardless of what direction they happen to be pointing in. This is an example of the additional "obvious" kind of principle I said we would be using in addition to our two basic principles. (It is called the principle of rotational invariance.)

to show that this is impossible; for if it happened, we could look at the same experiment from the point of view of an observer moving with stick B. This observer must also find that as the sticks pass, the 100-centimeter and zero-centimeter points of B coincide with the 95-centimeter and 5-centimeter points of A.* But for this observer stick A is moving while

> * This is the first place where we are using the principle that all observers agree on whether two things happen at the same point of space and time. In this case there are two such coincidences. One is the presence of the zero-centimeter mark of B and the 5-centimeter mark of A at the same point of space at the same moment of time; the other is the coincidence in space and time of the 100-centimeter mark of B and the 95-centimeter mark of A. Since the first observer (stationary with respect to stick A) said these two coincidences occurred, the second observer (stationary with respect to stick B) must agree that they occurred. As is always the case with the invariance of coincidences, we can convince ourselves more concretely that it must be so by imagining that some observable thing happens at the moment of coincidence. For example, here there could be a little piece of chalk attached to the 100-centimeter mark of B which leaves a white mark on the 95-centimeter mark of A as the two sticks pass. Since stick A can later be examined to check where the chalk mark is, the observer moving with stick B will have to admit that at some time the 100-centimeter mark of B did indeed pass by the 95-centimeter mark of A.

stick B stands still. He will therefore conclude that a stick moving perpendicular to its length (for him, stick A) gets *longer* than a stationary stick (for him, stick B). But the first observer concluded that a stick moving perpendicular to its length (for him, stick B) gets *shorter* than a stationary stick (for him, stick A).

Each observer concludes that stick B is shorter than stick

A, which appears to be perfectly consistent. However the principle of relativity has thereby been violated. The first observer did an experiment to find out what happens to the length of a stick moving perpendicular to itself and found that it shrank, and the second observer did an experiment* to find out what

> * It happens to be exactly the same experiment as the first man did, but this sharing of the same experiment need not bother us. Each man can regard it as his own private experiment, and simply note that the other man is watching.

happens to the length of a stick moving perpendicular to itself and found that it stretched. They did not come up with the same answer, as the principle of relativity says they must. Therefore the first man could not have found that the moving stick shrank.

In the same way the first man could not have found that the moving stick stretched, for if he had, the second man would have found that the moving stick shrank, and the principle of relativity would again have been violated.

The only possible outcome of the first man's measurement that does not lead to a contradiction of the principle of relativity is therefore that the moving stick has the same length as the stationary stick. Only in this case will the second observer reach the same conclusion.

We have therefore established with some effort a rule that is not terribly surprising. We shall call it Rule 1.

Rule 1.   A meter stick moving with uniform velocity in a direction perpendicular to itself has the same length as a meter stick at rest.

5
MOVING CLOCKS (I)

Having established that the length of a meter stick does not change when it moves* perpendicular to itself, we can use

> * From now on I shall stop adding the phrase "with uniform velocity" all the time. Whenever I talk about anything moving I shall always mean with uniform velocity with respect to inertial observers, unless I explicitly state otherwise.

this fact to construct a special kind of clock to investigate how fast moving clocks run compared with stationary ones.* Take

> * I shall also stop providing footnotes giving the expanded, correct versions of colloquial statements. I shall assume from now on that you understand without my reminding you that, for example, the last phrase in the text really should say ". . . how fast a given inertial observer finds a clock runs when it moves past him with uniform velocity and he measures its rate in terms of clocks that are stationary with respect to himself."

a meter stick, put a mirror at each end, and let a beam of light flash back and forth between the two mirrors. (We can make something more reasonable by getting more complicated, i.e., by having the return beam trigger off a device that sends out a new beam, so that the light does not die out after a while, but the basic mechanism is contained in the idealized version of a beam of light bouncing back and forth between two mirrors.)

We measure a time interval by counting how many times the beam of light goes back and fourth; i.e., every time the light hits the bottom mirror the clock "ticks." This is, of course, a rather peculiar clock. Once we have figured out how it works when moving, we shall consider all other clocks.

In Fig. 5.1 we see such a light-beam clock standing still. If we stand still next to the clock, we can analyze its operation quite easily. We first note that the speed of light is c. Therefore if l is the length of the meter stick, it takes a time l/c for the light to go from the bottom mirror to the top* and another

* Distance l = velocity c times time T; thus $T = l/c$.

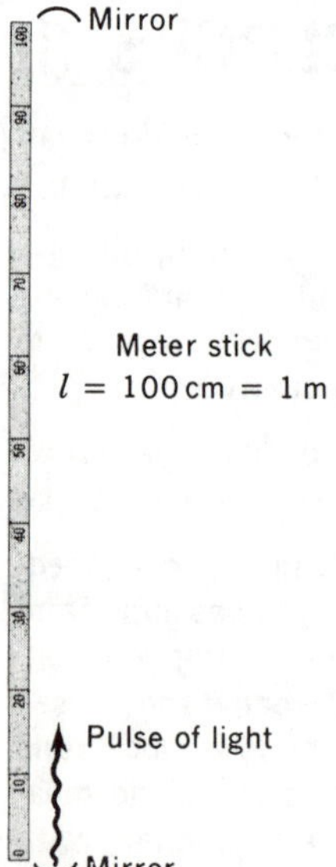

⌒ Mirror

Meter stick
$l = 100\,\text{cm} = 1\,\text{m}$

↑ Pulse of light

Mirror

Fig. 5.1 A simple light-beam clock at rest.

l/c for it to go from the top mirror back to the bottom one. Thus a complete round trip takes a time $t = 2l/c$, and this is the time interval between ticks of the clock, when it is stationary.

Now let us consider exactly the same clock, except that we shall suppose that it moves by us with some velocity v in a direction perpendicular to the length of the meter stick. How long does it now take between ticks? Let us call the time between ticks of the *moving* clock t'. The light spends half that time going from the bottom mirror to the top mirror and the other half going from the top back to the bottom. Now, how-

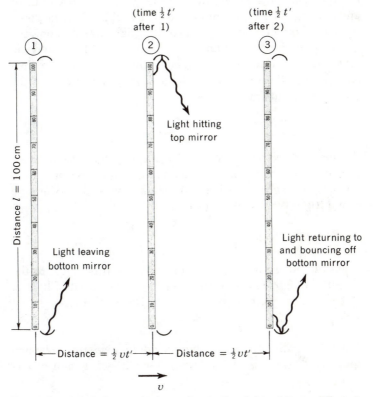

(time $\frac{1}{2}t'$
after 1)

(time $\frac{1}{2}t'$
after 2)

① ② ③

Light hitting
top mirror

Distance $l = 100$ cm

Light leaving
bottom mirror

Light returning to
and bouncing off
bottom mirror

|← Distance $= \frac{1}{2}vt'$ →|← Distance $= \frac{1}{2}vt'$ →|

v

Fig. 5.2 A light-beam clock moving to the right, at three different moments.

ever (see Fig. 5.2) the clock moves to the side during the time the light travels from the bottom to the top mirror, and so the light has to go a longer distance than it did when the clock was standing still. Therefore the moving clock takes a longer time between ticks.

We can calculate precisely how much longer a time it takes by calculating how much longer a path the light has to travel. We know that the up-down distance is still l, because we have seen in Chap. 4 that the length of a meter stick moving perpendicular to itself is the same as its length when stationary. The right-left distance the light has to go while traveling from the bottom to the top mirror is the velocity of the stick v times the time elapsed between the light leaving the bottom mirror and arriving at the top mirror, which is $t'/2$. Thus in going from the bottom to the top mirror, the light travels along the hypotenuse of a right triangle, one side of which is l and the other side of which is $vt'/2$. The Pythagorean theorem then tells us that the distance the light covers in going from the bottom to the top mirror is

$$\sqrt{l^2 + \left(\frac{vt'}{2}\right)^2}.$$

By the same reasoning we find that it covers the same distance in going from the top mirror back to the bottom mirror, and so the total distance (call it d) the light covers in making one round trip is twice this distance; thus

$$d = 2\sqrt{l^2 + \left(\frac{vt'}{2}\right)^2}. \tag{5.1}$$

On the other hand, this distance must also be equal to the time t' the light has been traveling times the speed of light c. But if $d = ct'$, we can conclude from Eq. (5.1) that

$$\tfrac{1}{2}ct' = \sqrt{l^2 + (\tfrac{1}{2}vt')^2}. \tag{5.2}$$

Equation (5.2) can easily be solved to give the unknown quantity t' in terms of the known quantities l, v, and c. The result is

$$t' = \frac{2l}{c\sqrt{1 - v^2/c^2}}. \tag{5.3}$$

Since we also know that the time t between ticks of the stationary clock is just $2l/c$,* we can rewrite Eq. (5.3) as an

* Note that the principle of the constancy of the velocity of light enters the argument in the assumption that the c appearing here is the same as the c appearing in (5.3).

equation relating the time t' between ticks of the moving clock to the time t between ticks of the stationary clock: *

$$t' = \frac{t}{\sqrt{1 - v^2/c^2}}. \qquad (5.4)$$

* Note that this result only makes sense if v is smaller than c. If v were larger than c, Eqs. (5.3) and (5.4) would contain the square root of a negative quantity. This is a mathematical clue that something has gone wrong, and indeed, if we go back and reexamine the argument assuming that v exceeds c, the trouble becomes evident: If the light-beam clock moves to the side with a speed greater than the speed of light, a pulse of light leaving one mirror is not going fast enough to catch up with the other mirror, and the clock never ticks at all. This same trouble will also appear in the next two chapters: The results of these chapters only make sense if none of the velocities involved exceeds the speed of light. We shall impose this limitation on the velocities to avoid this kind of trouble. It turns out to be no limitation at all. In Chap. 15 we shall see that no two material things can have a relative speed greater than the speed of light. It therefore does not matter that speeds in excess of c cause trouble in our analysis, since they turn out to be physically impossible to reach.

Since $\sqrt{1 - v^2/c^2}$ is smaller than 1, t' is bigger than t. Thus a light-beam clock takes a longer time between ticks when it is moving than when it stands still. This is the same as saying that a light-beam clock runs slower when it is moving than when it stands still.

One is tempted to conclude from the fact that a moving light-beam clock runs more slowly than a stationary one nothing more profound than that a light-beam clock is a very bad clock. This is wrong. The correct conclusion to reach is that any other conceivable clock, when moving, slows down in exactly the same way. For suppose we had, for example, a good Swiss watch. If we stand next to the light-beam clock with our watch, we can count the number of ticks the light-beam clock makes in 1 second as indicated by the watch. Since we know that a stationary light-beam clock takes a time $2l/c$ between ticks, this number will be $c/2l$. (This is because the total number of ticks given off in a second times the time between ticks must equal 1 second.)

Now consider an identical light-beam clock that moves away from us with velocity v (perpendicular to its length). Suppose somebody who is equipped with an identical copy of our Swiss watch moves along with the moving light-beam clock, so that he and his watch are also moving away from us with velocity v. This moving observer can count the number of ticks the moving light-beam clock gives in a time of 1 second as indicated by his moving watch. We know from the principle of relativity that he must get exactly the same answer, $c/2l$ ticks, as we got when we counted the number of ticks given by our stationary light-beam clock in a time of 1 second as indicated by our stationary watch. This is because *both of us have performed exactly the same experiment:* An experiment that measures how many ticks the same kind of light-beam clock gives during the time it takes a particular kind of watch, stationary with respect to the light-beam clock, to advance 1 second. The only difference is that his experiment is moving uniformly with respect to mine. But if that is the only difference, the principle of relativity says we must both come up with the same answer.

Now from our point of view, the moving light-beam clock is running slowly compared with our light-beam clock. Yet the moving observer tells us that while 1 second passes on his watch, the moving light-beam clock gives the same number of ticks as we found our stationary light-beam clock gave while 1 second passed on our stationary watch. The only way this can be true is if his moving watch has slowed down by exactly the same amount as the moving light-beam clock has slowed down.

Thus we are forced to conclude that a good Swiss watch, or any other kind of portable clock, slows down in the same way as a light-beam clock. This is Rule 2:

Rule 2.   If a clock takes a time *t* between ticks when it is stationary, then when it moves with velocity *v*, it takes a longer time $t/\sqrt{1 - v^2/c^2}$ between ticks. That is, when moving it runs at a slower rate.

It might appear that in extending the result for light-beam clocks to arbitrary clocks, we have used nothing but the principle of relativity. Actually we also used the principle of the constancy of the velocity of light, but it sneaked into the argument so quietly that you probably did not notice it. The second principle had to be used when we asserted that the moving man's light-beam clock was exactly the same as our light-beam clock. By this we meant that it was a standard meter stick with mirrors at either end between which a flash of light bounced back and forth with velocity *c*. If the moving man had found that the speed of the light in his light-beam clock was not *c*, the way we found that it was in ours, he would not have been performing the same experiment as we were, and nothing could have been concluded. But since the speed of light *is* *c* for all inertial observers, his light-beam clock must behave in exactly the same way for him as ours behaves for us, and our conclusion that his watch must also be running slower is correct.

You should convince yourself that there is nothing artificial or dishonest in the above argument. It is important to realize this, for the result is so unusual and yet so easily arrived at that, when seeing it for the first time, one invariably has the uncomfortable feeling that something tricky is going on. But if you go back and look over the argument, you will find that there was nothing up my sleeve. Nothing I said about light-beam clocks or watches, moving or standing still, was weird or grotesque. On the contrary, it was all rather commonplace and unimaginative. The fact that moving clocks run slower than stationary ones was demonstrated simply by applying the principles of relativity and the constancy of the velocity of light to a set of humdrum remarks and calculations. Once the (remarkable) principles and the (obvious) remarks are accepted, the (remarkable) conclusion must also be accepted.

Thus a moving clock runs slower than a stationary clock.

If you think about this for a while and also think about the principle of relativity, you will notice that something appears to be terribly wrong. For suppose we apply this result to a particular stationary clock A and a particular moving clock B. We have shown that B runs more slowly than A. But the principle of relativity tells us that we are also entitled to regard B as the stationary clock and A as the moving clock. If we had done this we should have had to conclude that, on the contrary, A runs more slowly than B.

These conclusions appear irreconcilable, and therefore the whole argument that moving clocks run more slowly is threatened by a reductio ad absurdum. However the apparent contradiction in the assertion that each clock runs more slowly, when measured in terms of the other, is illusory, and the conclusions are, in fact, consistent. To see this explicitly, it is necessary to use a few more properties of moving clocks and meter sticks, and we shall not be able to deal properly with this "paradox" until Chap. 10 (and again, in a somewhat different setting, in Chap. 16). In the present context the best I can do is to point out that the two conclusions about which clock runs slower do not necessarily contradict each other.

To see this we must drop the colloquial usage and return to the more precise, if clumsier, statements I used to call scrupulously to your attention in footnotes. When we say that a moving clock runs slower than a stationary one, what we really mean is this:

Any inertial observer will find that the rate of a clock moving past him with uniform velocity is less than the rate of a clock that is stationary with respect to him.

In the case of the two clocks A and B, I should have said that an observer who is stationary with respect to clock A will find that clock B runs slowly compared with clock A, and an observer who is stationary with respect to clock B will find that clock A runs slowly compared with clock B. This is certainly rather surprising, but it is not necessarily self-contradictory. It appeared to be self-contradictory when I stated it less carefully, saying baldly that a moving clock runs slower than a stationary clock, because in saying it that way I suppressed any explicit reference to the observer who reaches that conclusion,* thereby apparently implying that there was some

> * There was, however, an implicit reference. I had in mind
> that observer who is entitled to regard the first clock as sta-
> tionary. (If there had been no way at all of figuring out which
> observer I meant, I would have had no business putting it in
> the shorter, more colloquial language.)

absolute significance to such temporal judgments, independent
of all observers.

But in fact the more precise statement does not talk about
absolute temporal relations, valid for all observers. Rather, it
talks about conclusions reached by two different observers,
who examined various clocks and applied the laws of nature to
their findings. It is only our clinging to the old-fashioned idea
that statements about time have an absolute significance independent of who is making those statements that leads us to
feel that the conclusions of the two observers contradict each
other. Once we accept the idea (as we must) that time can
have different meanings for different observers, there need be
no contradiction between the two conclusions.*

> * It might be objected that one can, after a while bring the two
> clocks A and B right next to each other, and then, by com-
> paring them when they are face-to-face, check to see which
> one of them has advanced the most. When they are right in
> the same place at the same moment there is no possible
> ambiguity about this and it is certainly not possible that each
> has advanced less than the other. It must be that one of
> them—either A or B—has gained more time than the other,
> unless they have both gained exactly the same time.
>
> This is true, but at the moment, irrelevant. Our findings
> apply only to clocks that are moving uniformly. But if we wish
> to bring the two clocks back to the same place, at least one
> of them has to be turned around, and while it is being turned
> around, that clock is not moving with constant velocity (or,
> more technically, it is not in an inertial frame). Therefore we
> must be careful in reaching any conclusions. We shall ex-
> amine this question in detail in Chap. 16. Meanwhile note
> that it is not in conflict with anything said so far.

If you think this is nothing but a surrealistic intellectual
exercise, let me point out that the slowing down of moving

clocks has been observed and quantitatively verified. The clocks used are elementary particles or atomic nuclei, it being too difficult to accelerate a clock of macroscopic size to speeds close enough to c to make the effect appreciable.

Nuclei make good clocks because they can undergo internal vibrations of a precisely defined frequency, which can in turn be measured by measuring the frequency of electromagnetic radiation (gamma rays) given off in the course of these vibrations. If a piece of matter containing nuclei with a characteristic vibration frequency is heated, the frequency is found to decrease by a very small amount. This is because as the matter is heated, all the particles composing it, including the atoms to which the nuclear clocks belong, move back and forth with higher and higher velocities. The drop in the nuclear frequencies observed is just what one would expect by applying the relativistic slowing-down factor appropriate to the velocity increase associated with the temperature change.*

> * The effect is a very small one because the changes in velocity one can bring about by raising the temperature a reasonable amount are very small compared with c. It can be observed only because there exists a method (utilizing something called the Mössbauer effect) for measuring nuclear vibrational frequencies with extraordinary accuracy, thereby making it possible to detect minute changes in frequency.

Unstable particles give us clocks that can travel at speeds close to c. Mu mesons, for example, have the property that about half of any group of stationary mu mesons will decay (into an electron and two neutrinos) in about 3 microseconds. This time, characteristic of any mu meson, is a statistical property of the particle. Any particular mu meson may last longer or shorter than 3 microseconds, but on the average that is the length of time between its birth (through the decay of a pi meson) and death. High-speed mu mesons occur naturally in the form of cosmic rays, and with considerable ingenuity and effort, people have been able to measure the lifetimes associated with mu mesons traveling at various velocities. It is found that the faster the mu meson moves, the longer it lives on the average, the increase in lifetime being given by a factor $1/\sqrt{1 - (v/c)^2}$, where v is the speed of the meson.

APPENDIX
TO CHAPTER 5:
ANOTHER WAY
OF PROVING IT

The slowing down of moving clocks is so important a fact that it is worth proving in more than one way. The argument that follows makes no use of any properties of moving meter sticks. Let us call the factor by which a clock moving with speed v changes its rate, compared with a stationary one, g_v; that is, a clock which takes a time t between ticks when stationary will take a time $g_v t$ between ticks when moving with speed v. (In pre-relativistic days all would have agreed that g_v was just 1; we must prove that actually $g_v = 1/\sqrt{1 - v^2/c^2}$.)

Suppose, then, that I watch with a telescope a clock that when stationary takes a time t between ticks as it moves away from me with speed v. Let f_v be the slowing-down factor for the ticks I *see* through my telescope; i.e., a time $f_v t$ elapses between my *observations* of sucessive ticks. You might at first be tempted to say that f_v equals g_v but this would only be because you had forgotten to take into account (1) that between successive ticks the clock gets farther away and (2) that the light from the clock reaches my telescope after an interval of time that is longer for each successive tick, since the clock is continually getting farther away. The correct relation between f_v and g_v is easily worked out. Between two ticks of the clock a time $g_v t$ passes, during which the clock moves with velocity v and therefore travels a distance $v g_v t$. Hence between any two ticks the light has this much farther to travel before it reaches my telescope, and since its speed is c, it takes an additional time $v g_v t / c$ to get there. Hence the total time between my observation of the ticks is $g_v t + v g_v t / c$, and therefore $f_v t = g_v t + v g_v t / c$, or

$$f_v = g_v \left(1 + \frac{v}{c}\right). \tag{5.5}$$

Let us define f_{-v} in the same way as f_v except that the clock is moving *toward* us with velocity v. The relation corresponding to (5.5) is now

$$f_{-v} = g_v \left(1 - \frac{v}{c}\right), \tag{5.6}$$

the proof being exactly the same as it was for (5.5), except that now it takes $v g_v t$ *less* time for each successive flash to reach me, since the clock moves toward me.*

> * Since the slowing down of moving clocks depends only on the magnitude of their speed and not its direction (rotational invariance again), there is no need to define g_{-v}, since it is the same as g_v.

We now have two equations for three unknowns (f_v, f_{-v}, and g_v) and therefore need one more equation relating them. The equation we want is

$$f_v f_{-v} = 1, \tag{5.7}$$

which can be proved as follows:

Suppose I watch a very distant clock which is *stationary* with respect to me and which gives off a tick every t seconds. I shall certainly see a tick every t seconds because (1) the clock is stationary with respect to me and therefore ticks at its customary rate, and (2) it takes light from each tick the same time to reach me since the distance from me to the clock does not change between ticks. Now let an astronaut fly in a straight line with speed v from the distant clock to me, while I watch both him and the distant clock through my telescope. Suppose I observe N ticks of the distant clock from the moment I see him take off from the distant clock to the moment he lands on top of me. Now the astronaut has also been watching the distant clock and sees precisely the same number of ticks I do between takeoff and landing, since the light from each of these ticks has to go past his ship on its way from the distant clock to me. Every time the astronaut sees a tick he jumps for joy, and I, watching him, count the jumps. Since he jumps once for each tick he sees and he sees the same number of ticks I do, he will give N jumps during his journey; therefore I shall see him jumping at the same rate as I see the distant clock ticking: one jump observed every t seconds.* Now, how-

> * In fact my observation of his jumps will coincide with my observation of the ticks, for he jumps the moment the light from a tick reaches him, and immediately thereafter the light from the tick *and* the light from his jump travel on together in the same direction to me.

ever, we can also analyze the rate at which I see jumps this way: The astronaut can consider himself at rest and the distant clock as moving away from him with speed v. He therefore sees one tick every t_j seconds, where $t_j = f_v t$, according to the clock in his spaceship. I note, then, that the clock in his spaceship advances by t_j between each of his jumps, but since he is moving toward me with speed v, this means I see a jump every $f_{-v} t_j$ seconds, or every $f_{-v}(f_v t)$ seconds. But, as we said, I see a jump every t seconds, which means that f_{-v} times f_v must equal 1, so we have proved (5.7).

The rest is algebra:

Multiply the left and right sides of (5.5) and (5.6) together to get

$$f_v f_{-v} = g_v{}^2 \left(1 - \frac{v^2}{c^2}\right), \tag{5.8}$$

substitute (5.7) on the left of (5.8) to get

$$1 = g_v{}^2 \left(1 - \frac{v^2}{c^2}\right), \tag{5.9}$$

and solve (5.9) to find

$$g_v = \frac{1}{\sqrt{1 - v^2/c^2}},$$

which is what we set out to prove.

We have thus reached a conclusion about the rate of a moving clock identical to that arrived at by the apparently unrelated light-beam-clock construction. It is important to appreciate at this stage the fact that *any* reasonable argument that uses only the principle of relativity and the constancy of the speed of light will force us back to this same conclusion. We could go on in this way, producing more and more different kinds of arguments and thought experiments all leading back to the same conclusion. We would learn nothing more from this, but it would help to dispel the sense of uneasiness and suspicion that is invariably felt after seeing only one such argument. With only one argument offered, one tends to suspect that the conclusion depends on some trick in the argument— on things having been contrived in just the right way to lead to the desired conclusion. I hope two different arguments will begin to convince you that the conclusion has a validity that stands above any particular path leading to it.

6

LENGTH OF A
MOVING STICK (II)

In Chap. 3 we suggested that either moving clocks or moving meter sticks or both must behave in a peculiar way if the principle of the constancy of the velocity of light is right. In Chap. 5 we saw that moving clocks do indeed behave in a peculiar way, and in Chap. 4 we saw that nothing peculiar happens if the meter stick moves in a direction perpendicular to its length. In this chapter we shall see that something peculiar *does* happen if a meter stick moves parallel to its length.*

> * The question of what happens to a moving meter stick when it moves in a direction making some arbitrary angle θ with its length (Chap. 4 considered the case when this angle was 90 degrees, and this chapter considers the case when it is 0 degrees) is interesting but not particularly relevant to anything else we want to say. The answer is
>
> $$l' = l \sqrt{1 - v^2/c^2} / \sqrt{1 - \sin^2 \theta \, v^2/c^2}.$$

The technique we used to answer this question when the direction of motion was perpendicular to the stick is no longer helpful. Suppose we place a stationary stick parallel to the moving stick so the motion takes one stick right over the other. Because both sticks lie along the direction of motion, each point of one stick is at some time on top of every point of the other. Therefore it is hard to conclude anything at all from this. If, for instance, we again attach a bit of chalk to the 100-centimeter mark of the moving stick (stick B), the chalk will slide over the entire length of the stationary stick (A) and, instead of finding a single mark on stick A as we could do in the case discussed in Chap. 4, we shall find a single long white smear along the entire length of A. This does not help at all.

Apparently we have to do something more complicated. We could, for example, try to determine what point of the stationary stick was directly beneath the zero-centimeter mark of the moving stick at the time that the 100-centimeter marks of the two sticks were on top of each other. But to answer this question we have to know what we mean when we say that two events at different points in space happen at the same time. It turns out (as we shall see in Chap. 7) that we mean something rather more complicated than we would naïvely have thought, and so we shall avoid using a method that involves knowing what it means for two things in different places to happen at the same time.*

> * After we have learned what is meant by saying that two events in different places happen at the same time, we shall go back and measure the length of the moving stick this way as well (Chap. 9), to check that it gives the same result as the one we shall derive in this chapter.

Perhaps you think that I am just pigheadedly insisting on making a simple matter complicated. After all, you might say, if we simply move stick A along with stick B at the same speed, we can line them up next to each other and make a direct comparison at our leisure. But this certainly will not work, for we are trying to find out whether a moving stick shrinks, expands, or remains the same length as a stick at rest. If we try to do this by comparing the moving stick B with a stick A that is moving at the same speed, if a moving stick shrinks or expands, the stick A will shrink or expand by the same amount as B does. Since we do not yet know what this amount is, we cannot conclude anything at all by doing it that way.

So how can we do it? Here is one way: We can take the moving stick to be a light-beam clock, figure out how fast it ticks (remember, all we showed in Chap. 5 was how fast it ticks when it moves *perpendicular* to its length) without assuming anything about how long it is, and then figure out how long it has to be in order for it to tick at just that rate.

Therefore we shall answer two questions. The first is how fast a light-beam clock ticks when it moves, not perpendicular, but parallel to its length. Having answered that, we shall deduce from this rate of ticking how long the moving stick has to be.

The first question is easily answered by noticing that a light-beam clock remains a perfectly good clock whether it moves parallel or perpendicular to its length. But we saw in Chap. 5 that *any* moving clock slows down in the same way as a light-beam clock moving perpendicular to its length. In particular then, so will a light-beam clock moving parallel to its length. Therefore if the light-beam clock takes a time t between ticks when stationary and the meter stick of which it is made has a length l when at rest (that is, $l = 1$ meter), when it is moving, the time between ticks is

$$t' = \frac{2l}{c\sqrt{1 - v^2/c^2}} \tag{6.1}$$

(i.e., the same as Eq. (5.3)) regardless of whether the clock moves perpendicular or parallel to its length).

This argument may strike you as a bit too glib to be thoroughly convincing, but since the result is an essential one, you should be quite sure that you really believe it. The best way to convince yourself is to go back to the argument that any clock slows down in the same way as a light-beam clock moving perpendicular to its length and to repeat the argument for the special case in which the other clock is a light-beam clock moving parallel to its length.*

* Hereafter let us just say "perpendicular light-beam clock" and "parallel light-beam clock."

First consider two light-beam clocks at rest, with their meter sticks making a right angle and with the two zero-centimeter points in the same place (Fig. 6.1). Certainly these two clocks, being at rest, will tick at the same rate.* By this I mean that

> * Note that here again we are using the principle of rotational invariance. There should be no difference in the behavior of the two light-beam clocks if they differ only in that they point in different directions.

if two flashes leave the common zero-centimeter point at the same time,* one traveling up and one traveling to the side,

> * Here I can safely talk about two things happening "at the same time" since they also happen at the same place. I am using the principle of the invariance of coincidences. The two events are the up-going flash leaving the zero-centimeter mark and the sideways-going flash leaving the zero-centimeter mark. Since these two events happen in the same place *and*

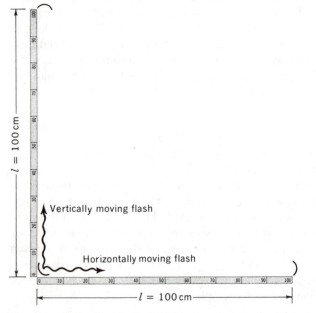

Fig. 6.1   Two identical light-beam clocks at right angles.

> at the same time, all observers will agree that they happen
> at the same place and time.

they will keep returning to the zero-centimeter mark at the
same time.

Now consider the same device moving uniformly with velocity v along a direction parallel to one of the sticks and perpendicular to the other. From the principle of relativity we
know that anybody moving along with the two clocks will
describe their behavior in exactly the same way we have, and
so he will also say that the two flashes keep returning to the
common zero-centimeter mark at the same time. But since all
observers agree on space-time coincidences, if he, moving with
the clocks, says the flashes keep arriving at the same place at
the same time, we, standing still, must also find that this is so.*

> * Here is an example of how in any particular application of
> the principle of the invariance of coincidences, we can
> imagine some sort of gadget that makes it obvious. We could
> attach a coincidence counter to the common zero-centimeter
> mark that would ring a bell only if it detected two flashes of
> light, one moving sideways and one moving vertically. Then
> either the bell would be ringing or it would not. If it kept
> ringing, everybody would certainly agree that it was ringing,
> regardless of how fast they might be moving.

But this can be so only if the perpendicular-moving flash continues to take the same time to complete each round trip as
does the parallel-moving flash, even when the two clocks move
past us, which means that the moving clocks continue to tick
together.

Having established how fast a parallel light-beam clock ticks
without having assumed anything about how long it is, let us
assume that when it passes us with velocity v parallel to its
length it has some unknown length l' and let us deduce what
l' must be in order for it to tick at the correct rate.

Figure 6.2 shows the moving light-beam clock, first at the
moment when the pulse of light leaves the left-hand mirror,
next when the light arrives at the right-hand mirror, and
finally when the light gets back to the left-hand mirror. (I
have displaced the three pictures vertically so they would not
get in each other's way, but of course the actual meter stick

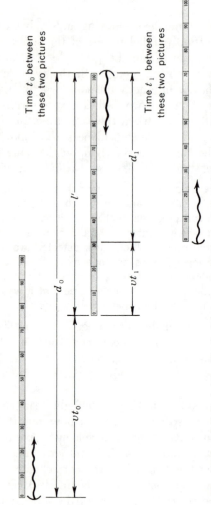

Fig. 6.2 A light-beam clock moving parallel to its length shown at three different moments. (The motion of the clock is entirely in a horizontal direction; the clock is displaced downward at successive moments in the figure only for the sake of legibility.)

moves on a single horizontal line.) We shall call the time interval* between the first two pictures t_0 and between the last

> * Let me remind you that what I really mean is the time interval as measured by an inertial observer past whom the stick is moving with uniform velocity v.

two t_1. The total time elapsed between the first and third picture is $t_0 + t_1$, which is just the time t' for a single tick.

We can find t_0 in terms of the unknown length l', v, and c in the following way:

In the time t_0, the light travels a certain distance which we shall call d_0. This distance is just the speed of light times the time elapsed, and since the speed of light is always c,

$$d_0 = ct_0. \tag{6.2}$$

On the other hand d_0 is the distance between where the left-hand mirror is in the first picture of Fig. 6.2 and where the right-hand mirror is in the second picture. In the first picture the right mirror is a distance l' away from the left mirror. Since a time t_0 passes between the first two pictures and the stick moves to the right with velocity v, at the time of the second picture the right-hand mirror has moved an additional distance vt_0 to the right; therefore the total distance d_0 is $l' + vt_0$ (see Fig. 6.2):

$$d_0 = l' + vt_0. \tag{6.3}$$

Equations (6.3) and (6.2) enable us to conclude that

$$ct_0 = l' + vt_0. \tag{6.4}$$

This can be solved to give t_0 in terms of l', v, and c:

$$t_0 = \frac{l'}{c - v}. \tag{6.5}$$

We can find t_1 in terms of the same three quantities by the same kind of reasoning:

Call the distance the light has to go between the times of the second and third pictures in Fig. 6.2 d_1. This is again the speed of light times the time elapsed; therefore

$$d_1 = ct_1. \tag{6.6}$$

Now d_1 is also the distance between where the right mirror is at the time of the second picture and where the left mirror is at the time of the third picture. At the time of the second picture the left mirror is a distance l' from the right, but during the interval t_1 between the two pictures it moves to the right a distance vt_1; thus (see Fig. 6.2):

$$d_1 = l' - vt_1. \tag{6.7}$$

Equations (6.7) and (6.6) tell us that

$$ct_1 = l' - vt_1, \tag{6.8}$$

and we can solve this for t_1 to find

$$t_1 = \frac{l'}{c + v}. \tag{6.9}$$

Finally, we know that t', the time it takes for the moving parallel light-beam clock to give one tick, is just the sum of t_0 and t_1. Adding the expressions given for t_0 and t_1 in Eqs. (6.9) and (6.5) we find that

$$t' = \frac{l'}{c - v} + \frac{l'}{c + v} = \frac{2l'}{c(1 - v^2/c^2)}. \tag{6.10}$$

This is the expression we need. It tells us what t' must be in terms of the unknown length l' of the moving light-beam clock, its velocity v, and the speed of light c. We can compare this with Eq. (6.1), which tells us what t' must be in terms of the same two velocities, and the length l of the same light-beam clock when it is stationary. In order for the two expressions for t', Eqs. (6.1) and (6.10), to be the same, it must be that

$$\frac{2l'}{c(1 - v^2/c^2)} = \frac{2l}{c\sqrt{1 - v^2/c^2}}. \tag{6.11}$$

This can be solved to give l' in terms of l:

$$l' = l\sqrt{1 - \frac{v^2}{c^2}}, \tag{6.12}$$

which gives us the length of a moving light-beam clock in terms of its length when standing still.

Since the length of a moving light-beam clock is just the length of the meter stick of which it is made, (6.12) tells us that a meter stick moving parallel to its length shrinks by a factor $\sqrt{1 - v^2/c^2}$ (which is always less than 1). This gives us Rule 3:

Rule 3. If a meter stick has a length l when it is stationary, then when it moves with uniform velocity v in a direction parallel to its length, it contracts to the length l', where $l' = l\sqrt{1 - v^2/c^2}$.

This shrinking of moving things along their direction of motion is known as the Fitzgerald contraction.* The lengthier

> * It is also sometimes referred to as the Lorentz contraction. Fitzgerald and Lorentz had both suggested, a few years before the theory of relativity appeared, that moving objects would contract in this way. However they did not arrive at this conclusion in anything like the way we have. They thought the contraction had to do with the nature of the electrical forces holding together the little particles of matter making up the body; if these forces became stronger in the right way, they could cause the body to shrink when it moved. This point of view is not wrong, but quite misleading. See Chap. 19 for further remarks on what "causes" the contraction.

noncolloquial statement of this effect is:

Any inertial observer will find that the length of a meter stick moving past him in a direction parallel to its length with uniform velocity v is less than the length of a meter stick that is stationary with respect to him by a factor $\sqrt{1 - v^2/c^2}$.

The same kind of remarks made at the end of the last chapter are in order here too. You should convince yourself that we reached this conclusion using only commonplace, reasonable, unobjectionable arguments. The algebra should not obscure that. There was no trickery involved. Any other sensible, routine, consistent argument would have led to the same conclusion. Having accepted the fact that moving clocks run slower than stationary ones, we are forced to conclude that moving meter sticks are shorter than stationary ones. It may be a surprising result, but given everything we have said in the first five chapters, it is unavoidable.

The problem described at the end of Chap. 5 arises here too. If we call the stationary meter stick A and the moving one B, an observer who is standing still with respect to A finds that B is shorter than A. But the principle of relativity permits us to consider the same two sticks from the point of view of an observer moving with stick B. For him, B is the stationary stick and A is the moving stick; therefore he will conclude that A is shorter than B. Thus depending on how you look at the same two sticks, you can find that either is shorter than the other.

As in the case of moving clocks, this appears self-contradictory only when one incorrectly assumes an absolute significance for something that does not have one. At the end of Chap. 5 we had to conclude that time can have a different meaning for different observers, and now we must conclude the same thing about length. We still have not discussed quite enough to give an explicit, step-by-step explanation of precisely how it is that each observer can conclude that the other man's stick is shorter than his own. At the moment we only know that it must be true and that our intuitive reason for thinking it incomprehensible and inconsistent is wrong. In Chap. 9 we shall return to this question and show that it can be simply explained.

Before continuing, it is useful to make a few definitions that enable us to speak without excessive awkwardness about lengths and time intervals in a way that is consistent with the facts we have learned about moving clocks and meter sticks. In pre-relativistic days one could talk without any ambiguity about the length of a meter stick, since it was thought that a meter stick had the same length whether moving or at rest. Now we know this to be false. Yet one feels that the particular length 1 meter does have something to do with a meter stick, which it indeed does, being, quite precisely, the length the meter stick will be found to have when measured by somebody who is at rest with respect to the meter stick.

The length of any uniformly moving thing as measured by somebody with respect to whom the thing is at rest is such a useful concept that it is given a special name: the proper length. In exactly the same way, if two things move uniformly with the same speed in the same direction, we call the distance between them as measured by somebody with respect to whom they are both at rest, the proper distance between them.*

> * Warning: Do not let the word "proper" make you think that in some way this distance is better than the distance measured by anybody else. A word like "stationary" would probably be better than "proper" since it is strictly neutral on whether we especially like that particular measurement, but all physicists use the word "proper."

We also define the proper time interval between ticks of a clock to be the time between ticks of the clock as measured by an inertial observer with respect to whom the clock is at rest.

Finally, when talking about something that is moving uniformly, we shall say that its "proper frame of reference" (or proper laboratory) is the frame of reference (or laboratory) in which it is at rest.

APPENDIX
TO CHAPTER 6:
ANOTHER WAY
OF PROVING IT

By continuing with the line of thought started in the Appendix to Chap. 5, we can construct an alternative short proof that a stick moving in the direction of its length must shrink.

First note that having proved that

$$g_v = \frac{1}{\sqrt{1 - v^2/c^2}},$$

we can substitute this into (5.5) to find that

$$f_v = \sqrt{\frac{1 + v/c}{1 - v/c}};$$

i.e., if we watch a clock that takes a time t between ticks when at rest, as it recedes from us with uniform velocity v, we shall observe a tick once every

$$t\sqrt{\frac{1 + v/c}{1 - v/c}}$$

seconds. The shrinking of moving sticks follows directly from this fact, as is demonstrated by the following experiment:

I take a stick that has length l when at rest and attach to the front end of the stick a clock that takes a time t between ticks when at rest. I then let the stick moving parallel to its length with speed v pass directly over my head, and I count the number of ticks of the clock I see between the moment when the clock is directly overhead and the moment when the rear end of the stick is directly overhead. If l' is the length of the moving stick, it takes a time l'/v for the stick to pass completely by me. During this time I watch ticks of the clock (which we take to pass at eye level); since it takes a time

$$t \sqrt{\frac{1 + v/c}{1 - v/c}}$$

between my observation of ticks, I see

$$\frac{l'/v}{t \sqrt{\dfrac{1 + v/c}{1 - v/c}}} = \frac{l'}{vt} \sqrt{\frac{1 - v/c}{1 + v/c}} \tag{6.13}$$

ticks altogether.

Now somebody with respect to whom the stick is stationary will agree on the total number of ticks I have seen while the stick went past my eyes.* But he will analyze what has

> * I could, for instance, make a mark on the stick for each tick I saw. Later we could count the marks.

happened somewhat differently. He will say that the stick has a length l, and since I moved from one end of the stick to the other with speed v, it took a time l/v for me to travel the length of the stick. He will also say that the time t' between my observations of ticks was the time t between ticks of the clock (t because he is stationary with respect to the clock) plus the time it takes for the light from the clock to travel the extra distance vt' that I moved during the time t'; thus

$$t' = t + \frac{vt'}{c},$$

which gives

$$t' = \frac{t}{1 - v/c}.$$

Now if I have seen one tick every $t/(1 - v/c)$ seconds during l/v seconds, the total number of ticks I have seen must be

$$\frac{l}{vt}\left(1 - \frac{v}{c}\right). \tag{6.14}$$

For this to be the same as (6.13) it must be that

$$\frac{l}{vt}\left(1 - \frac{v}{c}\right) = \frac{l'}{vt}\sqrt{\frac{1 - v/c}{1 + v/c}},$$

which requires that

$$l' = l\left(1 - \frac{v}{c}\right)\sqrt{\frac{1 + v/c}{1 - v/c}} = l\sqrt{1 - \frac{v^2}{c^2}}.$$

7

MOVING CLOCKS (II)

At this point we have deduced from the principle of relativity and the principle of the constancy of the velocity of light that moving clocks run slowly and moving objects shrink in the direction of their motion, compared with the corresponding stationary things. These two facts are extremely important, but it is essential to understand a further fact about moving clocks, before one can assemble the pieces and fully grasp the theory of relativity.

The additional point arises when one asks whether two clocks in different places are synchronized. It turns out that the notion of the synchronization of distant* clocks is a relative one.

> * By "distant" I just mean separated in space from each other—not necessarily very far apart, either from each other or from somebody observing them.

If one inertial observer finds that two clocks in different places are synchronized, there will be others who conclude that the same two clocks are not synchronized. We must convince ourselves of this and find a quantitative way of describing it.

Suppose an observer A sets up an inertial laboratory containing two identical clocks placed a distance l apart. He checks that they are l apart by laying down meter sticks along a line joining the clocks. He then synchronizes the clocks; i.e., he satisfies himself that at the moment clock 1 read zero, clock 2 also read zero. One way he can do this is to put a flashbulb on top of clock 1 that goes off when clock 1 reads zero. The light from the flash will then travel a distance l at a speed c to clock 2, taking a time l/c to get there. Therefore if clock 2 also read zero when the flash left clock 1, it will read l/c when the flash reaches it. By checking that this actually happens, A can confirm that the clocks are synchronized.

We now adopt the point of view of another inertial observer B past whom A and his laboratory are moving with uniform velocity v in the direction pointing from clock 2 to clock 1 (Fig. 7.1). B has watched A measure the distance between the two clocks and confirm their synchronization to his own satisfaction. From what *he* has just seen, B will conclude that the two clocks are *not* synchronized:

First of all B will decide that the clocks are not a distance l apart, but only a distance $l' = l\sqrt{1 - v^2/c^2}$ apart, since each of the meter sticks with which A measured the distance was moving parallel to its length with velocity v and therefore was contracted by a factor $\sqrt{1 - v^2/c^2}$. Watching A's synchronization check, B will certainly agree that clock 1 read zero when the flashbulb went off (space-time coincidence) and clock 2 read l/c when the flash reached it (another space-time coincidence). However he will not agree that it took a time l/c for the flash to travel from 1 to 2. He will say it took a time t', where t' is the distance the light had to travel divided by the speed of light c. Now the distance the light had to travel (Fig. 7.1) is l' (the distance between clocks 1 and 2) minus vt' (the amount that clock 2 moved toward clock 1 in the time t'); therefore

$$t' = \frac{l' - vt'}{c}, \qquad (7.1)$$

which, since $l' = l\sqrt{1 - v^2/c^2}$, can be solved to give t' in terms of l, v, and c:

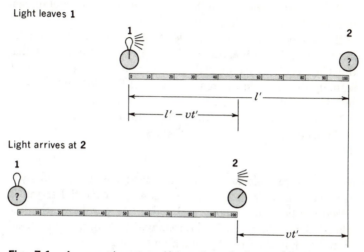

Fig. 7.1 An experiment verifying that clocks 1 and 2 are synchronized in their proper frame, as viewed by an observer moving with speed v from 1 to 2. The upper configuration occurs first, and the vertical displacement of the two pictures is for the sake of legibility.

$$t' = \frac{l}{c} \sqrt{\frac{1 - v/c}{1 + v/c}}. \qquad (7.2)$$

Now during the time t' that the light spent going from clock 1 to clock 2, each of the clocks, since it moved past B with velocity v, ran slowly and therefore did not advance by the amount t', but only by $t'\sqrt{1 - v^2/c^2}$. If we combine this with (7.2) we see that from B's point of view, during the passage of the flash from clock 1 to clock 2, each clock advanced by

$$\frac{l}{c} \sqrt{\frac{1 - v/c}{1 + v/c}} \sqrt{1 - \frac{v^2}{c^2}} = \frac{l}{c} - \frac{lv}{c^2}. \qquad (7.3)$$

Finally, since clock 2 read l/c when the flash reached it and advanced by $(l/c) - (lv/c^2)$ during the passage of the flash, it must have read lv/c^2 when the flash left clock 1. But clock 1 read zero when the flash left it. Therefore clock 2 is ahead of clock 1 by an amount lv/c^2.

B therefore extracts from his observation of A's synchronization check the following result:

If two clocks move with a velocity v parallel to the line joining them and an observer moving with those clocks says that the clocks are synchronized and separated by a distance l, actually the clock in the rear is ahead of the clock in the front by an amount lv/c^2.

We can abstract this still further to a more general, less prejudiced (B's word "actually" does prejudice things) statement:

Rule 4. If two clocks are synchronized and separated by a distance l in their proper frame, then in an inertial frame in which the two clocks move with velocity v parallel to the line joining them, the clock in the rear will be found to be ahead of the clock in the front by an amount lv/c^2.

It is therefore not enough to say that two clocks are synchronized, just as it is not enough to say that a single clock takes a certain time between ticks or that a meter stick has a certain length; all such statements only have meaning when the observer making the statement is specified.

Rule 4 is sometimes referred to as the relativity of the synchronization of moving clocks. It is also known as "the relativity of simultaneity." For if an observer A says that two clocks in different places are synchronized while another observer B says they are not, then if A says that two events occurring at those two places occurred at the same time, B will disagree. Indeed, if B moves with velocity v along the line joining the clocks, he can reason as follows:

A said the two events happened at the same time, which means that each of his clocks had the same reading, say zero, when the event next to it occurred. Since the clocks are stationary with respect to A, who says they are synchronized and separated by a distance l, the front clock is actually behind the rear clock by an amount lv/c^2. Therefore if the event in the rear happens when the rear clock says zero, the front clock still has to advance by lv/c^2 before it also says zero and the event in front occurs. Since the front clock moves with a speed v, it runs slowly and will take a time $(lv/c^2) / \sqrt{1 - v^2/c^2}$ to advance this amount, which means that the event in front will not happen until a time

$$\frac{lv}{c^2} \frac{1}{\sqrt{1 - v^2/c^2}}$$

after the event in the rear.

Thus Rule 4 can also be stated this way:

If two events are found in their proper frame to occur simultaneously and to be separated by a distance l, then in an inertial frame moving with velocity v parallel to the line joining the two events, the forward event will occur a time

$$\frac{lv}{c^2} \frac{1}{\sqrt{1 - v^2/c^2}}$$

after the event in the rear occurs.

All the results of this chapter apply in the case of observers moving parallel to the line joining the two clocks (or events). Suppose, instead, that an observer moved perpendicular to the line joining two clocks that were synchronized in their rest frame. In that case he would agree that they were synchronized. The easiest way to see this is simply to notice that if he thought one was ahead of the other, there would be no way of specifying which it was (as opposed to the parallel case, where it is the one in the forward direction that lags behind the one in the rear).*

* This is actually another application of rotational invariance.

If you do not like that argument (but it is perfectly correct), think of an experiment an observer stationary with respect to the clocks might do to verify their synchronization. He could have each clock emit a flash when it read zero and observe that the two flashes met at the midpoint of the line joining the two clocks. Now an observer moving perpendicular to the line joining the two clocks will see the two flashes leave the two clocks, and, after covering the same distance, meet. Since the speed of each flash is c and each covered the same distance, he will conclude that the flashes left their clocks at the same time, and so he will agree that the clocks are syn-chronized.

This gives us a rather uninteresting Rule 5:

Rule 5. If two clocks are synchronized and separated by a distance l in their proper frame, then in an inertial frame in which the two clocks move with velocity v perpendicular to the line joining them, the two clocks are also synchronized.

The alternative statement of Rule 5 in terms of simultaneous events is evidently:

If two events are found in their proper frame to occur simul-taneously and be separated by a distance l, in an inertial frame moving with velocity v perpendicular to the line joining the two events, the two events will also be found to occur simul-taneously.

APPENDIX
TO CHAPTER 7:
ANOTHER WAY
OF PROVING IT

To deduce that two clocks synchronized in their proper frame are not synchronized in a frame moving past them, we analyzed a synchronization check performed with light signals from the point of view of the moving frame. Now a simpler way to synchronize two spatially separated clocks is to bring them to the same place, check directly that they agree, and then carry one of them a distance l from the other. We must, of course, be very careful in concluding that the clocks are still synchronized at the end of this operation, for the clock that has been carried will have run slowly by virtue of its motion and will therefore be somewhat behind the clock with which it was originally synchronized.

To see how much of an error this introduces, suppose the second clock is carried the distance l from the first at a speed u, taking, therefore, a time l/u to get there. During this time it runs slowly, and therefore while the first clock advances by the full time l/u, the second only advances by $(l/u) \sqrt{1 - (u/c)^2}$ and is thus behind the first by an amount

$$\frac{l}{u}\left(1 - \sqrt{1 - \left(\frac{u}{c}\right)^2}\right) \tag{7.4}$$

when it reaches the full separation l (at which moment it comes to rest and remains behind by this amount thereafter).

Multiplying numerator and denominator of (7.4) by

$$1 + \sqrt{1 - \frac{u^2}{c^2}},$$

we can rewrite it as

$$\frac{lu}{c^2} \frac{1}{1 + \sqrt{1 - u^2/c^2}}. \tag{7.5}$$

But with the time difference written in this form it is clear that it can be made as small as one wishes by taking u sufficiently small, i.e., by carrying the clock sufficiently slowly. In other words if somebody specifies how far apart the clocks are to be and to within what fraction of a second they are to be synchronized, then by carrying the second clock away from the first sufficiently slowly, the effect of its running slowly during the trip can always be made small enough to maintain the desired accuracy.*

> * Note that this conclusion is due to two effects, one of which helps and one of which does not. The first is that the amount by which a moving clock slows down becomes very small when it moves slowly; the second is that the slower a clock moves, the longer it takes to cover a given distance, and therefore the more time it has to run slowly. Our analysis has shown that the first effect more than compensates for the second.

This is therefore a good method for synchronizing clocks, which has a certain old-fashioned directness and simplicity, since it makes use of no light signals. Nevertheless a moving observer watching this procedure will still conclude that the clock in front is behind the clock in the rear after the process is finished, for the amusing reason that for the moving observer, the effect of the tiny additional speed given to the second clock is not insignificant.

Suppose the observer moves with speed v along the line joining the clocks in a direction opposite to the motion of the second clock. He will then see the first clock moving (say) to the right with speed v and the second clock moving to the right with a slightly greater speed w while it is being carried away from the first.*

> * Non-relativistically w would just be $u + v$. We shall not assume this (because in fact it is not true—see Chap. 14) since all we shall need to know about w is that when $u = 0$, that is, when the two clocks have no relative velocity in their proper frame, then $w = v$, that is, the two clocks have no relative velocity in the moving frame.

The moving observer now calculates as follows:

The second clock moves with speed w until it gets a distance

$$l' = l \sqrt{1 - \frac{v^2}{c^2}} \qquad \text{(Fitzgerald contraction)} \qquad (7.6)$$

away from the first clock. The total distance the second clock covers in doing this is wt, where t is the time it takes. During the time t, the first clock has moved a distance vt, so the total distance the second clock has to cover is $vt + l'$. This tells us that

$$wt = vt + l'$$

or

$$t = \frac{l'}{w - v}. \qquad (7.7)$$

Now during this time t, the first clock runs slowly by a factor $\sqrt{1 - v/^2 c^2}$, while the second clock runs somewhat more slowly, by a factor $\sqrt{1 - w/^2 c^2}$. Hence when the process is completed and both clocks move with velocity v again, the second clock ends up behind the first by an amount

$$t \left(\sqrt{1 - \frac{v^2}{c^2}} - \sqrt{1 - \frac{w^2}{c^2}} \right) = \frac{l \sqrt{1 - v^2/c^2}}{w - v}$$

$$\left(\sqrt{1 - \frac{v^2}{c^2}} - \sqrt{1 - \frac{w^2}{c^2}} \right), \qquad (7.8)$$

(the right half of (7.8) coming from substituting (7.7) and then (7.6) in the left half). If we multiply numerator and denominator of (7.8) by

$$\sqrt{1 - \frac{v^2}{c^2}} + \sqrt{1 - \frac{w^2}{c^2}},$$

we find that another way of writing the amount by which the second clock trails the first is

$$\frac{l}{w - v} \left(\frac{w^2}{c^2} - \frac{v^2}{c^2} \right) \frac{\sqrt{1 - v^2/c^2}}{\sqrt{1 - v^2/c^2} + \sqrt{1 - w^2/c^2}}$$

$$= \frac{l}{c^2} (w + v) \frac{\sqrt{1 - v^2/c^2}}{\sqrt{1 - v^2/c^2} + \sqrt{1 - w^2/c^2}}. \tag{7.9}$$

Now the observer with respect to whom both clocks end up stationary maintains that as u gets closer and closer to zero, the accuracy with which the clocks are synchronized gets better and better. However as u approaches 0, w approaches v, so that $w + v$ approaches $2v$, while the ratio of the expressions involving square roots in (7.9) approaches $1/2$. Hence when u is very close to 0, (7.9) is very close to

$$\frac{lv}{c^2}. \tag{7.10}$$

Therefore the more accurately the stationary observer tries to synchronize the two clocks, the more precisely the moving observer will find the forward clock to be lv/c^2 behind the clock in the rear.

Finally note that this derivation of Rule 4 gives us an easy way of remembering that it is the clock in front that lags behind the clock in the rear, rather than vice versa. For if the two clocks start off moving together and synchronized and one is then moved ahead of the other, its velocity will be the greater of the two while it is being moved ahead, and hence it will slow down more. Conversely if one of the two is moved to the rear, from the point of view of the moving observer it is still moving forward, but more slowly than the clock to which nothing is done. Hence it will not run slowly by as much, and when it reaches the desired distance, it will be ahead of the forward clock.

8

OBJECTIONS AND REFLECTIONS

We have now found that in order for the principles of relativity and the constancy of the velocity of light not to be violated, any inertial observer must find that clocks and meter sticks moving past him with uniform velocity behave in the following ways: *

> * By the value of a quantity in the proper frame of some object let me remind you that we mean the value that quantity is given by an observer with respect to whom that object is at rest.

Rule 1. A stick moving with velocity v along a line perpendicular to its length has a length equal to its proper length.

Rule 2. The time between ticks of a clock moving with velocity v is longer than the time between ticks of an identical clock at rest by a factor $1/\sqrt{1 - v^2/c^2}$; that is, in a given length of time as measured by clocks at rest, the moving clock advances by only a fraction $\sqrt{1 - v^2/c^2}$ of that length of time.

Rule 3. A stick moving with velocity v along a line parallel to its length has a length equal to $\sqrt{1 - v^2/c^2}$ times its proper length; i.e., it shrinks.

Rule 4.   If two clocks synchronized in their proper frame are moving with velocity v parallel to the line joining them, then the clock in the rear is ahead of the clock in the front by an amount lv/c^2, where l is the distance between the clocks in their proper frame.*

> * Note that Rule 3, the contraction of moving sticks, is in a sense the spatial version of the temporal Rule 2, the slowing down of moving clocks. There is also a spatial version of Rule 4, the relativity of simultaneity, but it is usually not given explicitly, since it does not have the peculiarly relativistic aspects of the other three. If we state Rule 4 in its alternative form, "Two events which are simultaneous and separated by a distance l in their proper frame occur a time $(lv/c^2) / \sqrt{1 - v^2/c^2}$ apart in a frame in which they move with uniform velocity v," then the spatial version of this rule (which might be called the relativity of simullocality) reads, "Two events which occur in the same place a time t apart in their proper frame are separated by a distance $tv/ \sqrt{1 - v^2/c^2}$ in a frame in which they move with uniform velocity v." This is no more than a statement that by going to a moving frame, we can make stationary objects move with speed v (plus the slowing down of moving clocks, which is responsible for the factor $\sqrt{1 - v^2/c^2}$ in the denominator). To prove this, let the two events be the indications of two times t_1 and t_2 on a stationary clock (so $t = t_1 - t_2$). In a moving frame the clock runs slowly; therefore the time between the two events will be $t/ \sqrt{1 - v^2/c^2}$. During this time the clock moves with velocity v; so it covers a distance $vt/ \sqrt{1 - v^2/c^2}$ between the two events.

Rule 5.   If two clocks synchronized in their proper frame are moving with velocity v perpendicular to the line joining them, then the clocks are still synchronized.

A skeptic, examining the way in which these rules were found, might summarize our efforts up to now as follows:

We started off with an improbable rule that the speed of light is always c for any uniformly moving observer. In Chap.

5 this rule immediately got us into trouble, which we were able to avoid only by asserting that moving clocks run slowly. In Chap. 6 we were on the verge of showing that the slowing down of moving clocks leads to a contradiction, but again we escaped trouble by asserting that, in addition, moving meter sticks shrink. Then, in Chap. 7, we were well on our way to showing that both these assertions lead to nonsensical conclusions, but again escaped by asserting that events which are simultaneous for one observer need not be simultaneous for another. How much longer shall we keep this up? Must we invent newer and more bizarre rules each time the rules we already have lead us deeper into trouble?

There are two comments to address to such objections. First of all one has no grounds for objecting to any of the rules we have been forced to accept unless one either can demonstrate that one of the rules contradicts another, or can cite experiments which refute them. One may find the rules hard to accept psychologically, but as long as they are all mutually consistent, there is no logical ground for objecting, and as long as no experiment is found to contradict the rules, there is no empirical basis for rejecting them.

Furthermore we do not face an endless process of patching things up with successively stranger rules. The five rules we already have contain all the information we need. They are enough to give a complete, unique, and consistent description of the behavior of moving clocks and meter sticks.*

> * We shall, of course, find other rules describing the behavior of other things, but these rules all follow from the rules about clocks and meter sticks and require for their understanding no additional rules about clocks and meter sticks.

A persistent skeptic might take another line of attack:

If all these rules are correct and moving clocks and meter sticks behave so outrageously, why do we not notice such things happening about us?

The reason we do not is simply that c is an enormous speed compared with any we are likely to encounter in everyday life. Suppose we had an airplane that could go 30 kilometers per second, which is about 67,000 miles per hour—much faster than any plane that yet exists and considerably faster than the speed of a satellite in orbit about the earth. For such a speed v/c is still only one ten-thousandth. The quantity $\sqrt{1 - v^2/c^2}$, which gives the shrinking factor for lengths in the moving plane or the slowing down factor for clocks in the plane, is very nearly 1:

$$\sqrt{1 - \frac{v^2}{c^2}} = 0.999999995, \qquad \text{when } v = 30 \text{ km/sec.}$$

Thus if the plane were 100 meters long (somewhat longer than a football field) when on the ground, when in motion it would be 99.9999995 meters long if measured by somebody on the ground. It would thus lose 0.0000005 meter, which is about one fifty-thousandth of an inch of its length. The sheer technical difficulty of measuring the length of a moving plane from the ground to an accuracy of one fifty-thousandth of an inch is overwhelming. With the best instruments available today, such a shrinkage would be undetectable.*

> * Indeed, we should be hard pressed to measure the length of the plane with this accuracy even when it was in the airport and we could proceed at our leisure. Furthermore with the best equipment we could imagine, such a measurement would not be of much significance. Fluctuations in the length of the plane due to slight changes in the temperature of the air about it or due to all sorts of mechanical strains would be much larger than the tiny change in length due to the Fitzgerald contraction.

What about the slowing down of moving clocks? If we sat in such a plane for 1 hour (which is long enough for it to go almost three times around the world), our watches would lose 0.000000005 hour, which is about one sixty-thousandth of a second, compared with watches on the ground. If we wanted them to lose an entire second compared with clocks on the ground, we would have to stay aboard the plane for 60,000 hours, which is about 7 years. Furthermore if the plane only went at 0.3 kilometer per second (slightly faster than the speed of the fastest commercial planes), a watch would have to be aboard the plane almost 70,000 years before losing 1 second compared with watches on the ground.

This kind of consideration explains why we do not see the effects described by Rules 2, 3, and 4 (which are known as relativistic effects) all the time.

This does not mean, however, that these effects are of no practical interest. They are extremely important for the behavior of light (whose speed being c, is not small compared with c), they can be observed directly in the behavior of cosmic rays (which are elementary particles moving with speeds very close to c), they have consequences that it is essential to understand in designing accelerators in which various elementary particles (protons, electrons, etc.) are pushed up to speeds very near c, and they can be observed in the behavior of very distant astronomical objects, which are receding from us at speeds which are not minute compared with c.

I hope these remarks will quiet the objections. However I acknowledge that at this stage you may well feel quite uneasy about the five rules. It is not hard to accept the fact that they are correct for one particular inertial observer, but it appears to be very difficult to understand, in spite of our step-by-step derivation of the rules, how they can be true for all inertial observers. In particular you may be puzzled by questions like those I mentioned at the end of Chaps. 5 and 6:

Why does he say my sticks are shorter than his when I know that his are shorter than mine?

Why does he say my clocks run slower than his when I know that his run slower than mine?

Underlying these questions is a persistent, but (as we shall see) absurd question:

Which one of us is really right?

We shall answer these questions in the next two chapters. Meanwhile we shall not commit ourselves to who is "really right."*

* This is certainly prudent, since we know from the principle of relativity that the final answer will have to be that each is as right as the other.

However it is very hard to avoid committing ourselves to some observer's point of view, since the ordinary language we speak was developed by people who not only did not believe Rules 2, 3, and 4, but did believe incorrect versions of them.*

> * Incorrect Rule 2: The time between ticks of a clock moving with velocity v is the same as the time between ticks of an identical clock at rest; incorrect Rule 3: A stick moving with velocity v along a line parallel to itself has the same length as a stick at rest; incorrect Rule 4: If somebody finds that two clocks are synchronized, everybody else will also find they are synchronized.

It is therefore very dangerous to make any statements involving ideas of length or time without specifying how the lengths and times were measured. This does not mean we must always talk about light-beam clocks and flashbulbs; it is enough to specify which inertial observer made the measurements. For any given observer knows perfectly well what *he* means by his statements about lengths and times; he always has in mind quantities that he can measure using reliable clocks and accurate sticks. These clocks and meter sticks do not even have to be at rest with respect to him. If they are, his measurements are straightforward, but even if they are not, he can still make the measurements, as long as he remembers that the moving clocks and meter sticks behave as described in Rules 1 through 5. It happens that in this way he may be led to make statements that contradict statements made by other observers, but this will only happen if he uses words like "length" and "time" without qualifying them. If he is always careful to say "lengths as measured by me" and "times as measured by me," there will be, at least formally, no contradiction and certainly no ambiguity, because other observers will phrase their observations in terms of lengths as measured by them and times as measured by them.

Thus from now on we shall be wary of making any reference to lengths and times without specifying whose length and whose time. When considering two observers A and B, we shall talk about A-lengths, A-times, B-lengths, and B-times, by which we shall mean simply lengths and times as measured by A or B. Thus we shall only be talking about measurements, leaving open the question of what length and time really are.

9

WHY HE SAYS MY METER STICK SHRINKS ALTHOUGH I KNOW THAT HIS SHRINKS

We can now resolve the apparent paradox in the shrinking of a moving meter stick. Rule 3 is perhaps novel, but not distressing, if we only consider the shrinking of a meter stick moving by us compared with our own stationary meter stick. But the principle of relativity permits us to view the same situation with an observer going with the moving meter stick, and to him ours appears to be the shorter one.

This paradox disappears with the realization that a measurement of the length of a moving meter stick involves determining how far apart the two ends are *at the same time,* but since the ends are spatially separated, "at the same time" means different things to different observers (Rule 4). It is primarily the relativity of simultaneity that makes it intelligible that each of two observers should think the other's stick shorter than his own.

To see in detail how this works, consider two inertial observers A and B moving with velocity v with respect to each other. We shall design an experiment in which B measures a stick of A's and finds that it shrinks, and then examine B's experiment from A's point of view. We shall see that A not only maintains that his stick does not shrink, but has a perfectly reasonable and simple explanation for why B concluded that it did. Furthermore this explanation is completely consistent with A's knowledge that "actually" B's stick is the one that shrinks.

Let us suppose that A has a stick of proper length l, stationary with respect to A, and pointing along the direction of motion of B. B will say that the stick is moving past him parallel to its length with speed v. We therefore know from Rule 3 that if B measures the length of A's stick, he will find that it is only $l\sqrt{1-v^2/c^2}$. One way B could verify this is by taking a second stick of *proper* length $l\sqrt{1-v^2/c^2}$, stationary with respect to himself,* and aligning it parallel to the direc-

> * To avoid any possible misunderstanding, consider the particular case in which A's stick has a proper length of 100 centimeters (i.e., it is a standard meter stick) and v is numerically equal to three-fifths of the speed of light. (The value $3c/5$ is convenient because then $\sqrt{1-v^2/c^2}$ is simply $4/5$.) Then the stick B is using to convince himself that A's stick has undergone a Fitzgerald contraction has a proper length of 80 centimeters (i.e., four-fifths of 100 centimeters). B can make such a stick by sawing off the last 20 centimeters of a standard meter stick.

tion of motion of A's stick in such a way that A's stick slides over B's. (See Fig. 9.1.) B will then find that the left end of A's stick is directly on top of the left end of his own stick at the same time (B-time) as the right end of A's stick is directly over the right end of his own stick, which is convincing proof that A's stick has indeed shrunk to a length $\sqrt{1-v^2/c^2}$ meters.

The information B would enter in his notebook to describe the experiment would be:

(1)$_\text{B}$ My stick has a length $l\sqrt{1-v^2/c^2}$.*

> * B says the length of his stick is equal to its proper length, since the stick is stationary with respect to B.

(2)$_\text{B}$ When the clock on the left end of my stick reads zero, the left end of A's stick is right next to it.

A's stick

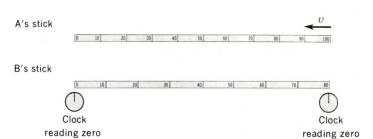

B's stick

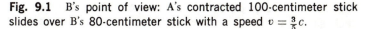

Clock
reading zero

Clock
reading zero

Fig. 9.1 B's point of view: A's contracted 100-centimeter stick slides over B's 80-centimeter stick with a speed $v = \frac{3}{5}c$.

$(3)_B$ When the clock on the right end of my stick reads zero, the right end of A's stick is right next to it.

$(4)_B$ The clocks on the two ends of my stick are synchronized; i.e., observations $(2)_B$ and $(3)_B$ were made at the same time.

$(5)_B$ A and his stick move with a speed v.

From observations $(2)_B$, $(3)_B$, and $(4)_B$, B will conclude that A's stick has the same length as his own, namely (from $(1)_B$), $l\sqrt{1 - v^2/c^2}$.

That is how B might verify the Fitzgerald contraction. If we give A B's notebook and ask him to comment on B's experiment, he will agree with observations $(2)_B$ and $(3)_B$ since they refer to coincidences in space and time.* When A looks

* In each observation the two things that occur at the same point of space and time are the pointing of a clock hand to zero and the presence of an end of a stick.

at observation $(1)_B$, he will note that B's stick moves past A with velocity v parallel to its length and, by applying Rule 3, will conclude that B's stick is shorter than B said it was by a factor $\sqrt{1 - v^2/c^2}$. Thus A will say that the length of the stick B used was actually only

$$l\sqrt{1 - \frac{v^2}{c^2}} \sqrt{1 - \frac{v^2}{c^2}} = l\left(1 - \frac{v^2}{c^2}\right) = 64 \text{ cm.*}$$

* Note that this by itself makes things even worse!

Finally, when A looks at observation $(4)_B$, he will note that

A's stick

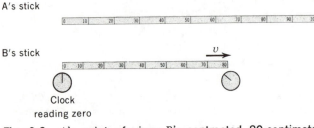

B's stick

Clock
reading zero

Fig. 9.2   A's point of view: B's contracted 80-centimeter
stick sliding under A's 100-centimeter stick with a speed
$v = \frac{3}{5}c$ at the moment of observation (2).

the clocks on B's stick are also moving past A with velocity v
parallel to the line joining them and that B's observation $(1)_B$
gives the proper distance between the clocks (since B moves
with the clocks). So A will apply Rule 4 to conclude that the
clock in the rear is ahead of the clock in front by an amount
$(l\sqrt{1 - v^2/c^2})\, v/c^2$. From this last conclusion he will say that
B's observations $(2)_B$ and $(3)_B$ were actually made at two
different times; so although B was able to draw a single picture (Fig. 9.1) depicting both observations, A will draw two
pictures, the first (Fig. 9.2) depicting the state of affairs at the
time (A-time) of observation $(2)_B$ and the second (Fig. 9.3)
depicting the state of affairs at the somewhat later (A-later)
time (A-time) of observation $(3)_B$. Note that in both pictures,
the clock in the rear is ahead of the clock in front and that
B's stick is shorter than A's.

A's stick

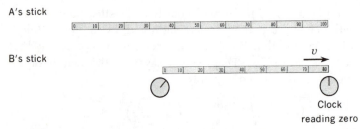

B's stick

Clock
reading zero

Fig. 9.3   A's point of view: B's contracted 80-centimeter stick
sliding under A's 100-centimeter stick with a speed $v = \frac{3}{5}c$ at the
moment of observation (3).

How much time (A-time) will A say has elapsed between observation $(2)_B$ and observation $(3)_B$? He can calculate that from Rule 2 and from the fact that the clock in front has to advance by an amount* $(l \sqrt{1 - v^2/c^2}) \, v/c^2$ between the two

> * It has to advance by this amount because at the time (A-time) of Fig. 9.2, it was behind the rear clock by this amount, and the rear clock said zero. At the time of Fig. 9.3 the clock in front says zero, and so it has indeed advanced by this amount.

observations. Since the clock moves, it runs slowly and therefore takes a time (A-time) $[(l \sqrt{1 - v^2/c^2}) \, v/c^2] / \sqrt{1 - v'^2 c^2}$ or simply lv/c^2 to advance by this amount.

Therefore by applying Rules 2, 3, and 4, A can correct the data in B's notebook and rewrite it in a version he considers to be correct. A's notebook would say:

$(1)_A$ B's stick has a length $l(1 - v^2/c^2)$.

$(2)_A$ When the clock on the left (rear) end of B's stick reads zero, the left end of my stick is right next to it.

$(3)_A$ When the clock on the right (front) end of B's stick reads zero, the right end of my stick is right next to it.

$(4)_A$ The clocks on the two ends of B's stick are not synchronized; observation $(3)_A$ occurred a time lv/c^2 after observation $(2)_A$.

$(5)_A$ B and his stick and clocks move with a speed v.

From the data in his own notebook, A can deduce how long his own stick is. The length (A-length) of A's stick is just the distance (A-distance) between the left end of B's stick at the time (A-time) of observation $(2)_A$ (Fig. 9.2) and the right end of B's stick at the time (A-time) of observation $(3)_A$ (Fig. 9.3). Between these observations B's stick moved to the right by an amount equal to its velocity v times the time (A-time) between the two observations lv/c^2, which is just $(lv/c^2)v$ or $l(v^2/c^2)$. Since B's stick moved by this amount between the two observations, to find the length (A-length) of A's stick, we have to add this distance to the length (A-length) of B's stick $l(1 - v^2/c^2)$. So A concludes that the length of his own stick is $l(1 - v^2/c^2) + l \, v^2/c^2 = l$, which is the right answer.

Now suppose for a moment that although A knew that he himself could use Rules 1 through 5, he knew nothing about the principle of relativity and considered himself to be absolutely at rest and his own conclusion to be the only correct one. A might then describe the experiment as follows:

"B's stick is the one that really shrinks. B is a fine experimentalist and made very accurate measurements, but he forgot the very important fact that he was moving with uniform speed. Therefore he failed to realize that his clocks were running slowly and out of synchronization and that his measuring stick had shrunk. As a result of these oversights, B erroneously concluded from his data that the length of my stick was only $l\sqrt{1 - v^2/c^2}$. If he had realized that he was moving, he would have applied the corrections I have applied to his data and reached the correct conclusion that the length of my stick is l.

"But as a result of overlooking his own motion, he did a very foolish thing. Although he thought he was directly comparing my stick with his, he actually concluded that my stick had the same length as a stick of his, the left end of which was over the left end of my stick at one time, the right end of which was over the right end of my stick at a later time, and which moved between those two times. Furthermore he failed to realize that his stick was shorter than he said and that his clocks were running slowly. Considering all this, it is hardly surprising that he incorrectly concluded that the length of my stick was $l\sqrt{1 - v^2/c^2}$."

A is perfectly entitled to hold this point of view. However B is also entitled to adopt the same point of view toward A, i.e., to conclude that A draws incorrect conclusions because *he* is unaware of *his* state of motion. If B did not know the principle of relativity and thought that he was at absolute rest and alone entitled to use Rules 1 through 5, B would say that A's stick really did shrink as B's measurement correctly indicated. He would go on to say:

"The reason A thought his stick had length l is that A forgot his stick was moving and consequently forgot to apply Rule 3 to get the correct answer, $l\sqrt{1 - v^2/c^2}$. The long and tortured argument A used to conclude from my data that his stick really has a length l, although, perhaps, a brilliant tour de force, is actually complete nonsense because I am the one who is really at rest; therefore my clocks, which are also at rest, really are synchronized and running at the right rate, and my stick, which is also at rest, is not shrunk at all. It was only by forgetting that I, B, am the only one entitled to use Rules 1 through 5 and incorrectly using them himself to describe my apparatus that A was able to twist my data into a confirmation of his erroneous view that his stick has length l."

Which of them are we to believe? To answer this we list those things both A and B *do* agree on. They both agree on observations (2) and (3). They would also both agree that if B's stick (which has a proper length $l\sqrt{1 - v^2/c^2}$) is laid on top of A's stick (which has a proper length l) when the two are at rest with respect to each other, B's stick would cover a fraction $\sqrt{1 - v^2/c^2}$ of A's stick. Finally, if B had checked that his clocks were synchronized by noting that light from a flashbulb halfway between them reached each of them just as each read zero, A, watching this, would have agreed that light from the bulb, halfway between, reached each as each read zero.*

* From which he would conclude, as we saw in a very similar case in Chap. 7, that B's clocks were *not* synchronized.

But taken together these facts, which both A and B agree on, are enough to give a complete description of B's experiment. A and B would agree on *all* the experimental data, if it were put this way:

(1) B's stick has a proper length $l\sqrt{1 - v^2/c^2}$ and B is at rest with respect to his stick.

(2) When the clock on the left end of B's stick reads zero, the left end of A's stick is right next to it.

(3) When the clock on the right end of B's stick reads zero, the clock on the right end of A's stick is right next to it.

(4) The clocks on the ends of B's stick have been synchronized in their proper frame and are at rest with respect to B.

(5) A and B move by each other with a speed v.

Observations $(1)_B$ through $(5)_B$ are simply translations of

(1) through (5) into the language B likes to use in describing things, and observations (1)$_A$ through (5)$_A$ are translations of the same facts into the language A prefers. Therefore the entire controversy between A and B is an argument not over facts, but over who uses the better language to describe these facts. The correct answer is that either language gives a clear and unambiguous description of the facts, as long as it is specified which language is being used. Therefore it is idle to speculate about who is really right. You can take your choice.

If, after reading only Chap. 1, on the principle of relativity, one were told that A insisted that observation (5)$_A$ was right and observation (5)$_B$ was wrong and B insisted that (5)$_B$ was right and (5)$_A$ was wrong and each denounced the other for taking the stand he did, one would see at once that the only wrongheaded thing about either was his having such an argument. What we have now seen is that the controversy over whether A's stick shrinks and B's does not or B's stick shrinks and A's does not is of precisely the same nature. It is easier to see the absurdity in the controversy over velocities because we are used to a statement about velocity only having a definite meaning when it is specified with respect to whom or what the velocity is being measured. Depending on whom we pick, either answer can be right. Once we realize that a statement about length also only has a definite meaning when it is specified whose clocks and meter sticks are being regarded as giving the "correct" length and time measurements, the controversy over whose sticks shrink becomes equally absurd.

10

WHY HE SAYS MY CLOCKS GO SLOWLY ALTHOUGH I KNOW THAT HIS GO SLOWLY

We can similarly dispose of the paradox raised at the end of Chap. 5. B notices that A's clocks are running slowly. The experiment he does to determine this is depicted from his point of view in Fig. 10.1 and 10.2. At time zero (B-time) A's clock is directly under a clock of B's and also reads zero (Fig. 10.1). A little while later (Fig. 10.2) it is under a clock a distance l (B-distance) away from the first clock. Since A's clock moves with velocity v, it has taken a B-time l/v for A's clocks to get to the position of Fig. 10.2; thus l/v is the time B's clocks read in Fig. 10.2. On the other hand since A's clock has been moving and therefore (Rule 2) running slowly, it reads only

$$\frac{l}{v} \sqrt{1 - \frac{v^2}{c^2}} \qquad (10.1)$$

in Fig. 10.2.

Now consider this from A's point of view (Figs. 10.3 and

B's clocks

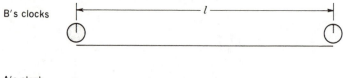

A's clock

v

Fig. 10.1   B's point of view: A's clock moving with speed v in the direction of the arrow past two of B's clocks. As the first B-clock is passed, A's clock agrees with it.

10.4). The first thing to notice is that B's clocks are not a distance l apart, but only a distance

$$l \sqrt{1 - \frac{v^2}{c^2}}, \tag{10.2}$$

since the stick joining them has shrunk. At time zero (A-time) A sees the configuration in Fig. 10.3. Note that A says that B's right clock is ahead of the left one by an amount lv/c^2, since it is the rear of two clocks synchronized and separated by a distance l in their proper frame (Rule 4). Since B moves past A with velocity v, a time (A-time)

$$\frac{l\sqrt{1 - v^2/c^2}}{v} \tag{10.3}$$

elapses between Figs. 10.3 and 10.4. During that time, B's

B's clocks

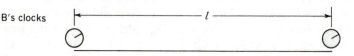

A's clock

v

Fig. 10.2   B's point of view: As the second B-clock is passed, A's clock lags behind it.

B's clocks

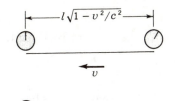

A's clock

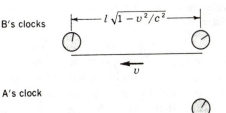

Fig. 10.3

B's clocks

A's clock

Fig. 10.4

Fig. 10.3   A's point of view: Two of B's clocks move with speed v in the direction of the arrow past A's clock. As the first B-clock passes over, A's clock agrees with it. At this moment the second B-clock is considerably ahead of A's clock. **Fig. 10.4**   A's point of view: As the second B-clock passes over, it is still ahead of A's clock, but not by as much as it was in Fig. 10.3.

clocks have been running slowly (by virtue of Rule 2), so that each only gains

$$\left(\frac{l\sqrt{1-v^2/c^2}}{v}\right)\sqrt{1-\frac{v^2}{c^2}} = \frac{l}{v}\left(1-\frac{v^2}{c^2}\right).$$

Therefore in Fig. 10.4 B's right-hand clock reads

$$\frac{lv}{c^2} + \frac{l}{v}\left(1-\frac{v^2}{c^2}\right) = \frac{l}{v}. \tag{10.4}$$

Thus A concludes that in Fig. 10.4 his own clock reads the time that has elapsed between Figs. 10.3 and 10.4,

$$\frac{l}{v} \sqrt{1 - \frac{v^2}{c^2}} \, ,$$

while the clock of B's directly over it reads the time

$$\frac{l}{v}.$$

But this is *exactly* what B concluded.

As in Chap. 9, A and B agree on all experimental observations* and differ only in how they choose to express the ob-

> * 1. The left side of Figs. 10.1 and 10.3
>   2. The right side of Figs. 10.2 and 10.4
>   3. The relative velocity v between A and B
>   4. The proper distance l between B's clocks.

served coincidences. B says A's clock runs slowly, whereas A says B reached this conclusion only because he forgot that "actually" his two clocks were both running slowly, they were both out of synchronization, and the stick joining them was contracted. The relativist says that either interpretation is permissible, provided it is made clear which observer is speaking, since they both lead to the same experimental conclusions.

Both this and the preceding chapter should make it clear that apparent paradoxes arise only when one concentrates on one of the peculiar properties of moving bodies, forgetting about the others. Thus the Fitzgerald contraction appears paradoxical only if one remembers that moving sticks shrink, while forgetting that moving clocks slow down and are out of synchronization. The reciprocal slowing down of moving clocks seems inconsistent only if one remembers that moving clocks slow down, while forgetting that they get out of synchronization with each other and that distances between them shrink. In fact, all three relativistic effects are necessary to explain either case properly.*

> * For some reason it seems to be the relativity of simultaneity that neophytes forget most often. A reminder that moving observers will not agree to the simultaneity of two events a stationary man says occur at the same time is usually enough to dispose of most of the "paradoxes" beginners come up with.

11
A
RELATIVISTIC
TRAGICOMEDY

CAST OF CHARACTERS

A
Friend of A
B
G
Chorus of Relativists

CRUEL NATURE

A one-act relativistic tragicomedy set in otherwise empty space

A, surrounded by his clocks and meter sticks, is talking with his friend.

**Friend
of A:**   Tell me, good A, is it then truly so
        That you are in a state of perfect rest?

A: I am, sir. I move not. My state of rest
        Is true and absolute.

F: Is it then so
        Your meter sticks do span a meter's length?

A: Not one jot more nor less, sir, I confess,
        Provided they maintain their state of rest.

F: How much, sir, in an honest hour's good time
        Will these, your clocks, have measured on their dials?

A: Faith, sir, an honest hour! No more, no less,
        While they remain with me, at perfect rest.

F: And will each of your clocks, regardless of
        The distance 'twixt them, read the same true time
        Upon their dials, all in that sweet relation
        That does befit fine clocks: Synchronization?

A: This too is so (once more the truth you've guessed!)
        Of all my clocks that, with me, are at rest.

F: Your rhymes improve at couplet's grace's expense.

A: Blank verse is not my business. Get thee hence.
F: A thousand pardons, sir! I did but jest
         And did not think it would disturb your rest.
A: My rest is perfect, absolute, and true.
F: In that case, gentle A, do you maintain
         That clocks and meter sticks that pass you by
         With uniform velocity (say v)
         Fail to be synchronized, slow down, and shrink
         As it is written in Rules One through Five?
A: Just so good friend, just so. You speak the truth.

B now floats uniformly into view, seated in the center of an immense network of clocks and meter sticks.

F: Look you! Who comes now?
A: That is Mr. B,
         Approaching us with constant speed (say v).
         Look how his clocks do fail to synchronize,
         Take longer than a second to describe
         A second's passage, while his meter sticks
         Do shrink along th' direction of his motion,
         All in accordance with my lovely rules.*

             * Which in this book are called
             Rules One through Five.

F: Welcome most hearty, B, to A's domain.
B: Nay, warmest welcome to both you and A
         As you progress toward my ancestral home.
F: How fare your many clocks and meter sticks?
B: Now and fore'er, sir, they are just and true.
         My clocks are in harmonious synchronization
         And in a second's time do indicate
         The passage of a perfect passing second.
         My meter sticks extend one meter's length
         From end to end.
F: Hear you that, A?
A: I do.
         The man has lost his wits. He does not know
         That he it is who moves, while I stand still.

Ergo the knave is fully unaware
That all his clocks and meter sticks behave
As it is written in Rules One through Five,
Failing to keep true time and span true length
To that extent precise and mathematick
As do my rules require for one who moves
Past me with his velocity.

F: Poor fool!
But now, as he passes by you, he will see
By swift comparison, experimental,
Of his askew equipment with yours true,
That his is deep in error.

A: No, alack!
You overestimate the wisdom of
The man. So deep has he enmeshed himself
In folly, so fully does he deem himself
At rest, that he believes that my Rules One
Through Five describe the sticks and clocks at rest
With me!

F: A double folly's double woe!
But yet methinks there consolation be
In double error. The saving point is this:
That if to his false-deemed state of rest erroneous
He adds a further concept incorrect
And gross, by his most wrongful application
Of your Rules One through Five, which we both know
Describe the strange distortions of things moving
Past him who is at rest (and such are you),
If, as I say (for I have lost the thread
Of my intent) he wrongfully applies
Your special rules, assuming they are his,
Then marry, by this double error of his
(Wrongly to deem himself at rest, and worse,
Wrongly to think that he can use your Rules)
Does he not double the chance of contradiction
Which will his error correct, his mind inform,
When he observes your instruments of measure
So just and true (due to their state of rest)?

A: His second folly does abet his first

And by compounding, save it. Had he but thought
Himself at rest and not as well considered
My own Rules One through Five, his too to use,
His error, by th' impending confrontation
Of swift advancing B and my true tools
Of space and time, would manifest become
To B himself, forced to this recognition
By contradiction palpable and merciless.
Howe'er because he uses my own Rules
As if he were at rest, and I the mover,
Along with my true clocks and meter sticks,
The inconsistencies that should inform
His intellect of its sad misconception
And jar it like a ringing clarion call
To certain knowledge of those clear distortions
His many clocks and meter sticks are heir to
By virtue of their motion, he poor fool
Is able to account for in a way
That masks the inconsistencies and bars
Sweet ministering contradiction from
The portals of his mind. He simply blames
The facts that should destroy his sleep dogmatic
On the fictitious shrinkage, slowing down,
And lack of that sweet quality we deem
Most excellent in clocks: Synchronization,
That he in his most vain, deluding use
Of my Rules One through Five assigns to my
Most wrongfully malignéd instruments.
To his misfortune, Nature, that arch deceiver
So made the world that his delusions, two,
Will learn from this encounter nothing new.
Each doth the other in falsehood dark confirm.
So was it e'en with C and X and Y.
So shall it be when G and H come by.
I rage against such cruel deception vainly;
Harsh Nature has decreed it.

F: (*to B, now very close*)   Look you, sir
At the clocks and meter sticks of outragéd A!
See you that yours and his argue not the same?

B: Of course they differ: His meter sticks do shrink,
           His clocks are slow, nor are they synchronized;
           While my sticks measure distance absolute,
           My clocks record Time's true and even tread,
           Each, though apart, my other clocks do prove.
           This is because, quite simply, I don't move.
A: Alas, poor B! Nature conspires against him.
B: Alas, poor A! He thinks that I be mad,
           When all too well I know the madness lies
           Within him. So has it been with Y and C
           And X; with G and H, so shall it be.

B passes by A and recedes into the distance.

F: O wicked Nature! So to conspire 'gainst B
           That all his gross and lamentable follies
           Most undetectable thy tricks have rendered.
A: Sadder still, that but for delusions twain
           He hath a most incisive, cogent brain.
           Alas, poor B! And C! And X! and G!
B: (*from afar*) Alas, poor A! and X! and G! and C!
G: (*coming into sight*) Alas, poor A and B! And X!
               and C!
Chorus of
Relativists:
           Such sorry discord need not be
           If Absolutists had more sense:
           So right in all their measurements,
           So mad in their philosophy.

12

HOW MUCH IS PHYSICS AND HOW MUCH SIMPLE ARITHMETIC?

This chapter describes a completely non-relativistic model which displays all the baffling relativistic effects. Its purpose is as much to soothe as it is to instruct, since it illustrates nothing new. However seeing these effects occur in a setting of familiar, slowly moving objects may help you to lose any remaining sense of mystery or contradiction over how it is that each of two observers can find the other's clocks slower, his sticks shorter, etc.

In this non-relativistic model we shall consider two sets of clocks and meter sticks moving very slowly with respect to each other (so that no relativistic effects need be considered). One set consists of real meter sticks and real synchronized clocks, while the other consists of incorrect instruments: meter sticks on which the marks are deliberately drawn too near each other and clocks that are adjusted to run too slowly and that are set out of synchronization. The amusing thing about the model is that if I tell you everything you want to know about the two systems *except* which is the correct one, it is impossible to determine this; i.e., if you erroneously assume that the distorted system is correct and the correct one distorted, you will conclude that the correct system is distorted in precisely the way that actually the incorrect system is distorted.

We shall avoid any possibility of genuine relativistic effects confusing the issue by considering only velocities less than 10 meters per second—so small a fraction of the speed of light that no relativistic effects could be detected. We can therefore forget about relativity and, for the small velocities to be considered, conclude safely that moving meter sticks have the same length as stationary ones and that moving systems of clocks remain synchronized and run at the same rate as stationary ones. If you think relativity has anything to do with the nature of the model we are about to examine, you will have missed the entire point, for what follows can be understood completely by anybody who is familiar with the ordinary (non-relativistic) properties of slowly moving ordinary clocks and meter sticks.

The model consists of a real meter stick with correct clocks and another incorrectly made meter stick with clocks that run too slowly and are out of synchronization. Both clock-stick systems are shown in Fig. 12.1 and, at a somewhat later time, in Fig. 12.2. The shaded strips with the numbered markings on them are supposed to be pictures of the two meter sticks. Stick A is a real meter stick, and hence (as is evident from the pictures) stick B is a shrunken stick; i.e., successive numbers are too close together on stick B. The rectangular boxes below stick A and above stick B are supposed to be clocks, the number in the box indicating the reading of the clock in seconds at the moment the picture (suppose each figure is a photograph of the two clock-stick systems) was taken. Since the clocks attached to stick A, the true meter stick, are correct clocks, the picture in Fig. 12.1 was taken at a time of 2 seconds, and that in Fig. 12.2, a second later, at 3 seconds. It is evident that the clocks attached to stick B are not synchronized, since at the instant the picture in Fig. 12.1 was taken, they had a whole range of readings from −0.4 to 3.0 seconds. You can see from Fig. 12.2 that the clocks attached to stick B also run slowly, for although the picture in Fig. 12.2 was taken

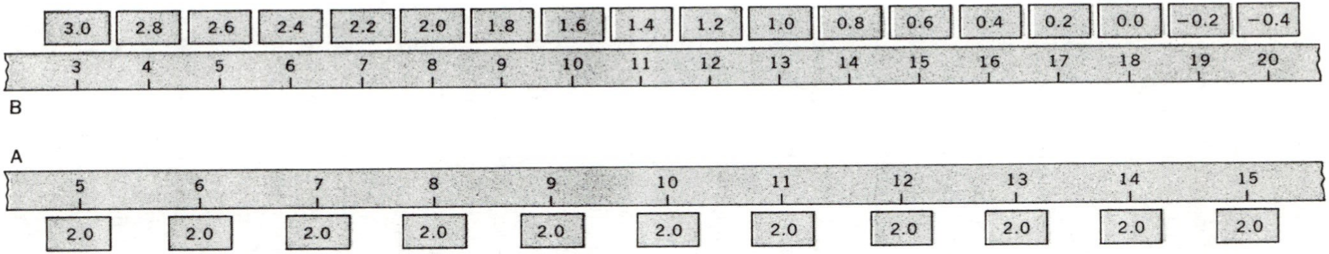

Fig. 12.1 The A and B clock-stick systems at time 2 seconds. The numbered marks on stick A are 1 meter apart, and the numbered marks on stick B are only 0.6 meter apart. System B moves to the right at 3.2 meters per second, and clocks on B get further behind by 0.2 second at each successive mark to the right.

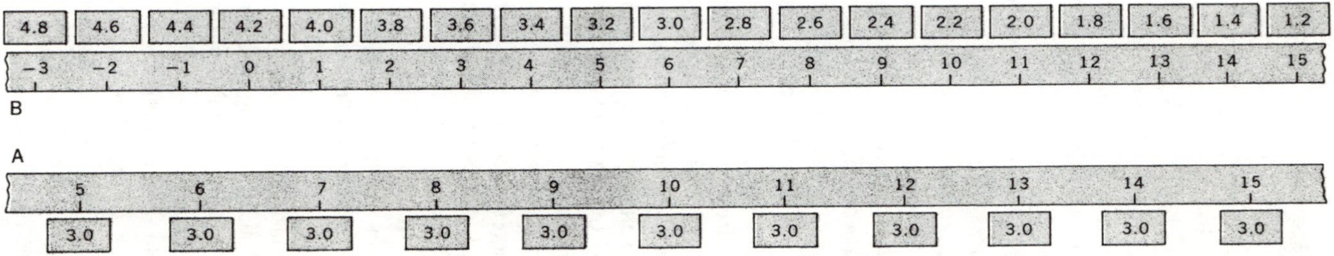

Fig. 12.2 The configuration of the A- and B-systems 1 second after Fig. 12.1. Stick A has not moved and all the clocks on A have advanced by 1 second. Stick B has moved 3.2 meters to the right and all its clocks have advanced by 0.6 second.

a second after that shown in Fig. 12.1, each clock attached to stick B has only advanced by 0.6 second. (The clock above mark 4 on stick B at 2 seconds (Fig. 12.1), for instance, reads 2.8, but a second later at 3 seconds (Fig. 12.2) it has advanced to 3.4, gaining only 0.6 second.) Note also that between the two figures, stick B has moved to the right by 3.2 meters (for example mark 8 on stick B has moved from directly above the 8-meter mark on A (Fig. 12.1) to directly above 11.2 meters (Fig. 12.2)). These two observations completely characterize the difference between Figs. 12.1 and 12.2:

In the second that has elapsed between the two pictures, stick B has moved 3.2 meters to the right, and every clock on stick B has advanced by 0.6 second.

The precise way in which the scale on stick B is contracted and the clocks on stick B are asynchronous and too slow is not arbitrary, but carefully chosen in the following way:

I first drew the two pictures of stick A appearing in Figs. 12.1 and 12.2, which represent that portion of a correct stationary meter stick extending from slightly below 5 to slightly above 15 meters, at two moments of time a second apart, as indicated by the clocks attached to stick A, each of which has advanced from 2 to 3 seconds between the two pictures.

Next, I picked two velocities, $v = 3.2$ meters per second and $c = 4$ meters per second. (They could have been picked at random, but I chose a pair that would lead to a picture that was fairly easy to follow.) With this particular choice of v and c, note that $v/c = 0.8$, $\sqrt{1 - v^2/c^2} = 0.6$, and $v/c^2 = 0.2$ second per meter.

Armed with these numbers, I drew system B, which is a picture of a bad meter stick with bad clocks which mimic all the relativistic effects. Such a system could have been made by first taking a long, unmarked piece of meter-stick wood and painting on it the numbers 3, 4, 5, 6, . . . at points carefully measured to be only 0.6 meter apart, thus mimicking the Fitzgerald contraction (remember $\sqrt{1 - v^2/c^2} = 0.6$). You can check from either of the figures that this is the way in which the scale on B is condensed. For example, in Fig. 12.1, the distance between marks 8 and 13 of stick B is just 3 meters, since they extend from the 8- to the 11-meter mark on stick A.

Hence five units of stick B extend 3 meters, which means that each one extends 3/5 meter, or 0.6 meter.

The next step in constructing system B was to manufacture a set of clocks all of which ran slowly by a factor of 0.6 ($= \sqrt{1 - v^2/c^2}$), thus mimicking the slowing down of moving clocks. That is why, as pointed out above, each clock on stick B has advanced only by 0.6 second in the second that elapsed between the two pictures.

Next, in setting up the slowly running clocks along stick B, I set them out of synchronization according to the rule that as one moved a distance l of stick B's units to the right, the clocks should get behind by an amount lv/c^2. Since v/c^2 is 0.2 second per meter, this means that for each B-unit that one moves to the right along stick B, the readings of the clocks should decrease by 0.2 second, which is clearly the case in both figures. (This mimics Rule 4. Remember that the l appearing in the relativistic formula lv/c^2 is the distance between the clocks in their proper frame. Since in our model the slow clocks move with stick B, their asynchronization should be proportional to the distance between them measured in the incorrect "B-meters.")

Finally, having painted the scaled-down meter stick, and attached the slowly running unsynchronized clocks to it, I set the whole apparatus B in motion to the right along stick A at a speed of 3.2 (v) meters per second. Hence a second later the picture in Fig. 12.2 shows stick B has moved 3.2 meters to the right.

At this point let me emphasize that system B is something anybody could make in his kitchen, with a long unpainted strip of meter-stick wood, a supply of black paint, a set of ordinary clocks, and a rudimentary knowledge of watchmaking (sufficient to enable him to make them slow down enough, which he could do, for instance, by adjusting the tension of the spring in the balance wheel). Indeed, if this were a lecture instead of a book, I would have made such a machine and you could examine it for yourself. As it is, you will have to be content with the two pictures of it shown in Fig. 12.1 and 12.2.

You should check carefully that these two pictures do depict precisely the situation I have described; i.e., by comparing the incorrect clock-stick system B with the correct system A, verify to your own satisfaction that stick B moves to the right at 3.2 meters per second, that its scale is condensed by a factor of 0.6, that its clocks run slowly by a factor of 0.6, and that the readings of the clocks on stick B drop behind by 0.2 second for each additional "B-meter" (i.e., each additional 0.6 real meter) one advances in the forward direction along B.

If you have convinced yourself of this, I must apologize, for actually I have misinformed you in one minor point. Everything I said about the two clock-stick systems is quite correct except that I mixed them up. I should have said that stick B is a genuine meter stick with correct synchronized clocks attached to it, and stick A moves to the left at 3.2 meters per second, has a scale that is shrunken by a factor 0.6, and has clocks that run slowly by a factor of 0.6 and drop behind by 0.2 second for each "A-meter" one advances in the forward direction.

Of course you know that I am joking, for a glance at the two photographs shown in Fig. 12.1 and 12.2 convinces one that system A is the right one. But I did not tell you how those two pictures were taken. Actually it *was* stick A that moved to the left. It does not look that way in the picture because the camera was firmly attached to stick A, with the region between the 5 and 15 marks on A within its field of view. The reason stick B appears to have moved to the right between the two pictures is simply that the camera, being attached to stick A, moved past stick B to the left.

Thus you cannot tell from the pictures which of the two systems is moving unless I tell you whether the camera moved or not. Does that shake your faith in the correctness of system A? Probably not, for you can still see clearly from the pictures that system A's clocks are synchronized and system B's clocks are not, and hence, if one of the systems is correct, it must be system A. But that assumes that all parts of each photograph were exposed at the same time. Actually they were not, because something was wrong with the camera.

The camera was an ordinary camera with the shutter jammed open; i.e., it was stuck on a long time exposure. To avoid overexposing the film and blurring the picture, an opaque

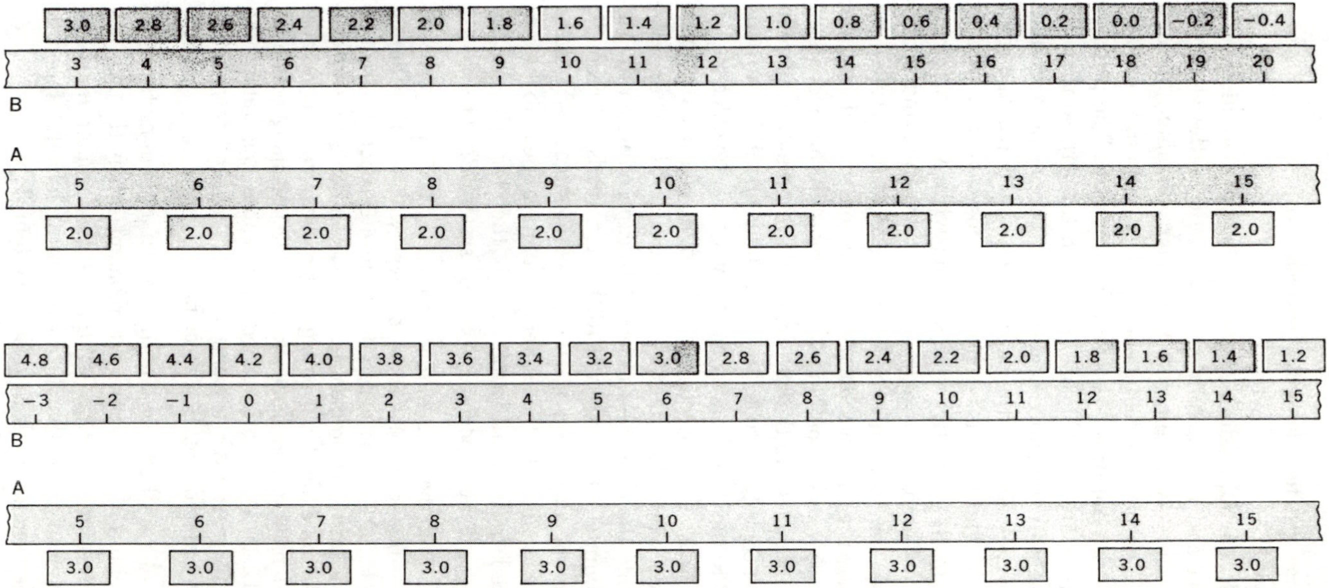

Figs. 12.1 and 12.2 Figs. 12.1 and 12.2 presented again for your contemplation. *Convince yourself thoroughly* that stick B moves to the right at 3.2 meters per second, that its marks are only 0.6 meter apart, and that its clocks only advance 0.6 second in a second and drop behind by 0.2 "B-seconds" for each "B-meter" one advances in the forward direction along stick B.

black cloth was hung directly in front of the two sticks, between them and the camera. Only the side of the cloth the sticks were on was illuminated, the camera remaining in total darkness. Now the cloth had a narrow vertical slit in it, initially off to the right, out of the camera's field of view. The pictures were taken by moving the black cloth to the left, so that the slit passed across the entire field of view of the camera, ending up again out of view on the extreme left. In its passage from right to left, the slit passed by and exposed to the camera every part of the two stick systems, and when working properly, it moves across so rapidly (say in a ten-thousandth of a second) that an essentially instantaneous exposure results.

However when the particular pictures you see in Figs. 12.1 and 12.2 were taken, something was wrong with the mechanism that pulled the cloth to the left, and it moved much too slowly to give an accurate picture. In fact it took about 3.4 seconds for the slit to pass all the way across, which is revealed in the fact (Fig. 12.1) that the clock on the right of stick B, which was exposed when the slit started moving across, reads −0.4 second, while the clock on the left of B, exposed when the slit was just completing its passage to the left, reads 3.0 seconds. The reason the clocks on B (which, with apologies, I again remind you is actually the correct system) appear to be unsynchronized in Figs. 12.1 and 12.2 is therefore due to the slow passage of the exposing slit across system B.* (Evidently

* Note that the time during which Fig. 12.1 was exposed (from −0.4 to 3.0 seconds) overlaps with the time during which Fig. 12.2 was exposed (1.2 to 4.8 seconds) creating a minor technical problem which you can settle to your own satisfaction either by inventing a second camera and suitable sets of mirrors and black cloths or by imagining that after Fig. 12.1 was photographed, the entire apparatus was stopped, reset back to some earlier time, and then set in motion again for Fig. 12.2 to be made.

0.2 second elapses between the time when a given meter mark on B is behind the slit and the moment when the slit has passed far enough to the left for the next meter mark on B to be exposed. Thus the slit moves to the left at a speed of 5 meters per second with respect to system B.)

So system B is actually the correct one, and the clocks on B in the two pictures appear to be unsynchronized for the simple reason that different parts of the picture were exposed at different times. Having accepted this fact, we can, by correctly interpreting the two pictures, confirm the fact that all the inaccuracies I originally attributed to system B are in fact properties of system A.

First, how fast does system A move to the left? We have to figure that out again, since our earlier observation that system B moved to the right at 3.2 meters per second was based on an analysis that assumed system A was right. We can get the correct answer by noting the location in terms of stick B of any particular point on stick A at two moments of time according to the clocks on B. For example (Fig. 12.3a), the 11-meter mark on A is opposite the 13-meter mark of stick B at a time of 1.0 second. (Remember that the clocks on B are the correct ones and the moving slit was vertical; thus to find the time at which any event happened, we must look at the clock on B directly above that event.) In Fig. 12.3b the 11 mark on A is now opposite the $7\frac{2}{3}$-meter mark on stick B, and the clock on B, if there, would read a time one-third of the way from the time indicated by the clock at 8 meters (2.6) to the time indicated by the clock at 7 meters (2.8),* that is,

* This is because the slit in the camera moves uniformly to the left.

$2.6 + \frac{1}{3}(0.2) = 2\frac{2}{3}$ seconds. Hence the 11-meter mark of A has moved $13 - 7\frac{2}{3} = 5\frac{1}{3}$ meters in $2\frac{2}{3} - 1 = 1\frac{2}{3}$ seconds. The speed of stick A is therefore $5\frac{1}{3}/1\frac{2}{3} = 16/5 = 3.2$ meters per second. Thus our conclusion about the relative velocity of systems A and B remains unchanged.

Next, let us find the true distance between successive numbers on stick A. The fact that they appear to be farther apart than on stick B in the pictures is misleading. For example, the part of stick A (11) under the 13-meter mark of stick B (Fig. 12.4) was photographed 0.2 second after the part of stick A (11.6) lying under the 14-meter mark. During that 0.2 second A moved 3.2 meters per sec × 0.2 second = 0.64 meter farther to the left, and hence A looks longer in relation to B than

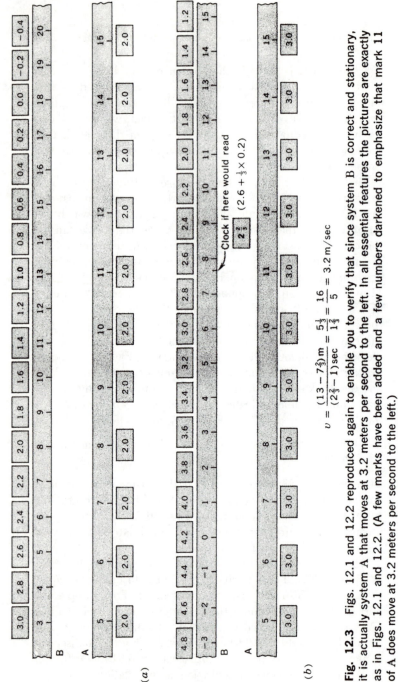

Fig. 12.3 Figs. 12.1 and 12.2 reproduced again to enable you to verify that since system B is correct and stationary, it is actually system A that moves at 3.2 meters per second to the left. In all essential features the pictures are exactly as in Figs. 12.1 and 12.2. (A few marks have been added and a few numbers darkened to emphasize that mark 11 of A does move at 3.2 meters per second to the left.)

$$v = \frac{(13 - 7\frac{4}{5})\,\text{m}}{(2\frac{2}{3} - 1)\,\text{sec}} = \frac{5\frac{1}{5}}{1\frac{2}{3}} = \frac{16}{5} = 3.2 \,\text{m/sec}$$

Clock if here would read
$2\frac{2}{3}$ $(2.6 + \frac{1}{3} \times 0.2)$

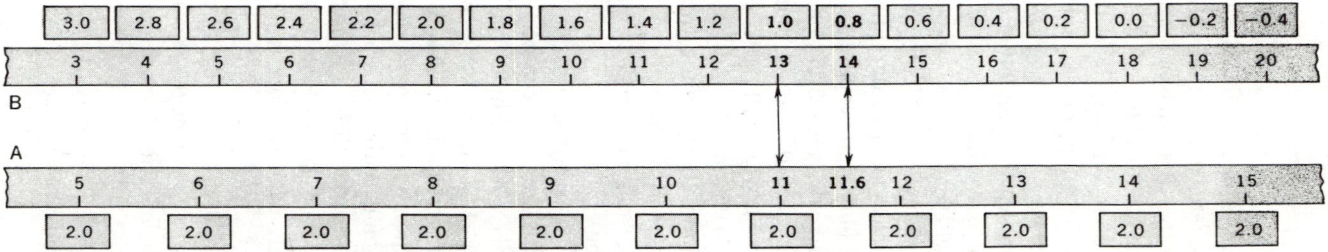

Fig. 12.4 Fig. 12.1 reproduced again to demonstrate that since system B is correct and stationary, it is actually stick A that has a shrunken scale. Mark 11 is not as far to the left of 11.6 as it appears, since it was photographed 0.2 second later, during which A moved $0.2 \times 3.2 = 0.64$ meter to the left. At the moment 11.6 was under mark 14 of B, 11 was thus under 13.64 of B, and the true distance between them only 0.36 meter.

it really is. Actually at the moment the 11.6-meter mark of stick A was under the 14-meter mark of stick B, the 11-meter mark of stick A was under the 13 + 0.64-meter mark of stick B. Hence a length of stick A equal to 11.6 − 11.0 = 0.6 "A-meter" is actually 14 − 13.64 = 0.36 meter long. Thus the "A-meter" is only a fraction 0.36/0.6 = 0.6 of a meter, and so it is really stick A that has shrunk by 0.6.

You should convince yourself that any other reasonable way of figuring out the length between divisions of stick A, given that system B is correct, will lead to the same conclusion. For example, we can piece together parts of the two photographs to get two pieces of the one correct photograph we might have made if all parts of the film were exposed simultaneously. Thus (Fig. 12.5) at 3.0 seconds, mark 5 of stick A was under the 3-meter mark of B and mark 10 of A was under the 6-meter mark of B. Hence a picture in which all parts of the film were exposed simultaneously at 3.0 seconds would show the part of stick A between marks 5 and 10 stretching from the 3-meter to the 6-meter marks of stick B. Hence 10 − 5 = 5 units of A are 6 − 3 = 3 meters long, and so a single unit of stick A is 3/5 meter, or 0.6 meter long.

We can also verify that it is in fact system A's clocks that are out of synchronization in precisely the way we incorrectly said system B's were. Take, for example, the clocks at the 5- and 10-meter marks of stick A. At 3.0 seconds (see Fig. 12.5) the clock at the 5-meter mark of stick A reads 2.0 seconds, while the clock at the 10-meter mark reads 3.0 seconds (thus coincidentally agreeing with the clock on B directly above it, which gives the correct time). Thus the clock at 5, which is 5 "A-meters" ahead of the clock at 10, is behind it by 1 second, which is precisely what the lv/c^2 rule tells us, since l is 5 while v/c^2 is 0.2.

Finally, we can verify that it is actually system A's clocks that are running slowly by a factor of 0.6. Look, for example (Fig. 12.6), at the clock below the 11 mark of stick A, which reads 2.0 seconds when the true time is 1.0 second and 3.0 seconds when the true time is $2\frac{2}{3}$ seconds. It has thus advanced 1 second in $1\frac{2}{3} = 5/3$ seconds; i.e., it runs slowly by a factor of $3/5 = 0.6$.

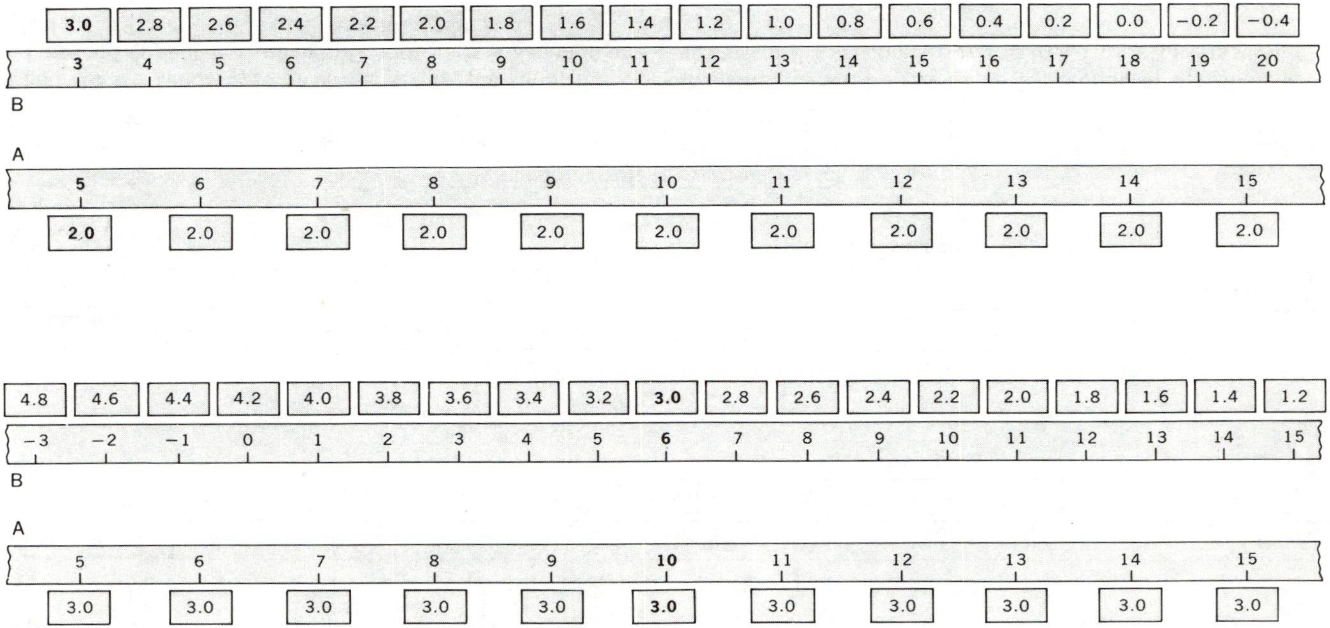

B

3.0	2.8	2.6	2.4	2.2	2.0	1.8	1.6	1.4	1.2	1.0	0.8	0.6	0.4	0.2	0.0	-0.2	-0.4
3	4	5	6	7	8	9	10	11	12	13	14	15	16	17	18	19	20

A

2.0	2.0	2.0	2.0	2.0	2.0	2.0	2.0	2.0	2.0	2.0
5	6	7	8	9	10	11	12	13	14	15

B

4.8	4.6	4.4	4.2	4.0	3.8	3.6	3.4	3.2	3.0	2.8	2.6	2.4	2.2	2.0	1.8	1.6	1.4	1.2
-3	-2	-1	0	1	2	3	4	5	6	7	8	9	10	11	12	13	14	15

A

3.0	3.0	3.0	3.0	3.0	3.0	3.0	3.0	3.0	3.0	3.0
5	6	7	8	9	10	11	12	13	14	15

Fig. 12.5 Since system B reads the true time and distance, at 3 seconds mark 5 of A was under mark 3 of B (upper part of figure) and mark 10 of A was under mark 6 of B (lower part of figure). The picture is identical to Figs. 12.1 and 12.2, except for the darkening of the numbers and marks referred to.

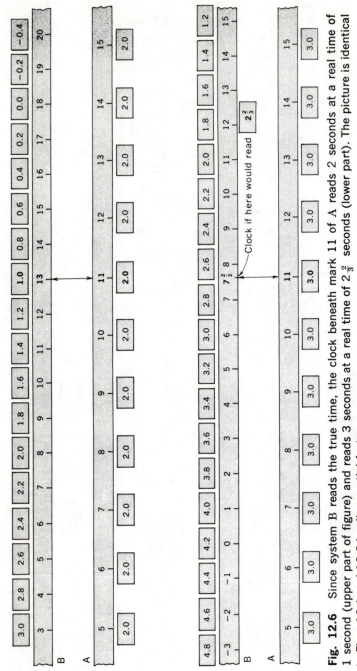

Fig. 12.6 Since system B reads the true time, the clock beneath mark 11 of A reads 2 seconds at a real time of 1 second (upper part of figure) and reads 3 seconds at a real time of $2\frac{2}{3}$ seconds (lower part). The picture is identical to Figs. 12.1 and 12.2 in all essential features.

Thus Figs. 12.1 and 12.2 have the property that if we are told that system A is correct, it follows that stick B moves 3.2 meters per second to the right relative to A, that stick B is too short by a factor of 0.6, and that the clocks on B run too slowly by a factor of 0.6 as well as reading 0.2 second farther behind for every "B-meter" one advances in the direction of stick B's motion. But if, on the contrary, we are told that system B is correct, it follows that stick A moves 3.2 meters per second to the left relative to B, that stick A is too short by a factor of 0.6, and that the clocks on A run too slowly by a factor of 0.6 as well as reading 0.2 second farther behind for every "A-meter" one advances in the direction of stick A's motion.

The possibility of this peculiar reciprocal situation in which if we consider one system correct we conclude that the other is contracted, running slowly, and out of synchronization has nothing at all to do with the *physics* of relativity, i.e., with the physical behavior of real objects moving at speeds close to 300,000 kilometers per second. This was deliberately emphasized by the fact that nothing in our model moved faster than a few meters per second and that *c* was just 4 meters per second. Once clocks and measuring sticks were invented, anybody could have noticed as a purely intellectual exercise that if one did construct a shrunken stick with appropriately asynchronized and slowly running clocks, then taking the incorrect system to be correct, would make the correct system seem incorrect in precisely the same way. The historical fact is that until experiments started to indicate that the speed of light was the same for all inertial observers, nobody did notice this curious property of spatial and temporal measurements. Prior to such experiments, there had been no particular motivation for examining how space-time measurements were made in moving frames of reference, in a way that might have led to the discovery of this reciprocity.

Suppose, however, that long before the advent of special relativity somebody had been clever enough to discover the two pictures shown in Figs. 12.1 and 12.2.* If he had believed

* If he were clever enough to figure out the relativistic rules, he would probably also have come up with another set of

rules which also work and which genuinely have nothing to do with the real world:

There is a velocity c such that moving sticks *get longer* by a factor $\sqrt{1 + v^2/c^2}$; moving clocks *go faster* by a factor $\sqrt{1 + v^2/c^2}$; and if two clocks are synchronized and separated by a distance l according to instruments moving with them, the clock in front is *ahead* of the clock in the rear by lv/c^2.

in the principle of relativity, having discovered the effects shown in these figures, he would have realized that there are many things moving clocks and meter sticks can do and remain consistent with the principle. They can, of course, remain unchanged. Another possibility, however, is that there is a special velocity c, such that clocks moving with speed v slow down by a factor $\sqrt{1 - v^2/c^2}$, meter sticks moving parallel to their length with speed v shrink by a factor $\sqrt{1 - v^2/c^2}$, etc. If this were so, the principle of relativity would not be violated, since anybody considering the moving system to be correct would reach the same conclusions about the stationary one.

Our pre-relativistic analyst might then have realized that his empirical knowledge was not enough to enable him to conclude that in our universe c is infinite (i.e., that moving clocks and meter sticks are unaffected by their motion), but only that c was very much larger than any velocity he had yet encountered. This would explain why none of the effects had been observed, and leave open the question of the actual value of c.

Now there are many ways of figuring out what the value of c actually is. Perhaps the simplest is this: Consider an object that moves at a speed of 4 meters per second (which is the value of c used in constructing the model) according to A (Fig. 12.7), so that, for example, it is at the 5-meter mark of stick A at 2.0 seconds and at the 9-meter mark of stick A at 3.0 seconds. That will ensure that its velocity is 4 meters per second if stick A is right. If, on the contrary, stick B is right, then since the object was at the 5-meter mark of A when the clock there read 2.0, it was also at the 3-meter mark of B when the clock there read 3.0 (Fig. 12.7). If it passed the 9-meter mark of A when the clock there read 3.0, it was also at the

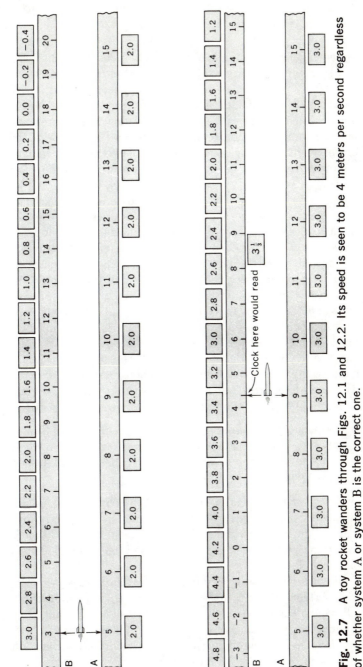

Fig. 12.7 A toy rocket wanders through Figs. 12.1 and 12.2. Its speed is seen to be 4 meters per second regardless of whether system A or system B is the correct one.

$4\frac{1}{3}$-meter mark of B when the clock there read $3\frac{1}{3}$ (Fig. 12.7). Hence according to system B, the velocity of the object was $4\frac{1}{3} - 3 = 1\frac{1}{3}$ meters in $3\frac{1}{3} - 3 = 1/3$ second; that is, $4/3$ meters in $1/3$ second, or, again, 4 meters per second. Thus an object which moves with speed c according to system A also has the same speed c according to system B. By finding an object which has the same velocity in two systems moving uniformly relative to one another, one can establish the value of c.

The entire physical content of that part of special relativity that deals with space-time measurements is contained in the single experimental fact that the actual value of the special speed c is 2.9979 hundred thousand kilometers per second, as demonstrated by the fact that light has this speed past any inertial observer. From the point of view of this chapter, the significance of the experimental measurement of the speed of light and the fact of its constancy for all inertial observers is that it singles out from the general class of ways in which moving clocks and meter sticks might behave the particular one characteristic of the physical world. The answer $c = 2.9979$ hundred thousand kilometers per second (instead of $c = \infty$) was considered revolutionary only because people had not noticed the other possibilities.* It is also clear from this point

* By $c = \infty$ we mean not that the velocity of light is infinite, but that the special velocity occurring in Rules 2 through 4 is infinite, i.e., that moving clocks and meter sticks behave according to pre-relativistic notions.

of view that the special thing about 2.9979 hundred thousand kilometers per second is *not* that it is the speed of light, but that, on the contrary, the special thing about light is that it is something whose speed is 2.9979 hundred thousand kilometers per second.

Of course if we really had in front of us the two clock-meter-stick systems pictured in Figs. 12.1 and 12.2, we could easily tell which was correct. The actual systems are not completely symmetrical, since one was constructed in a very peculiar way that involved painting meter numbers in unorthodox places on a blank meter stick and altering originally correct clocks. This is because the value we chose for c in our model was only 4 meters per second. If we wanted to make such a model for any value of c other than 2.9979 hundred thousand kilometers per second, we should have to alter one set of clocks and meter sticks. However, (and here is a statement about the nature of the physical world) in the special case in which $c = 2.9979$ hundred thousand kilometers per second, it really would be impossible to tell which clock-meter-stick system was correct, because not only would each indicate the other to be distorted in exactly the same way, but also the method by which each was constructed would be symmetrical.

For if I consider system A to be correct, my instructions to somebody else for building a system B for this special value of c are simply to take a set of ordinary meter sticks and ordinary clocks, set the meter sticks in motion parallel to their length with speed v, and, while moving with the stick, line up the clocks along the meter sticks and check that they are synchronized and running properly. On the other hand, somebody moving with and using system B would provide exactly the same instructions to me for setting up system A. For the special value $c = 2.9979$ hundred thousand kilometers per second, things are completely symmetrical, and it is genuinely impossible to tell which system is the right one.

13

THE LORENTZ
TRANSFORMATION

In this chapter we shall not derive any new physical effects from the two basic principles, as we did in Chaps. 4 to 7, nor shall we look at these effects in a manner that makes them more intelligible, as we tried to do in Chaps. 9 to 12. The point of the present chapter is only to introduce a concise and universally used method of summarizing all the relativistic effects contained in Rules 1 through 5, the Lorentz transformation equations.*

> * The transformation is named after Lorentz because he discovered it within the context of electromagnetic theory. It was not until Einstein, however, that it became clear how the Lorentz transformation equations were to be interpreted, and that their applicability transcended electromagnetism.

It is hardly necessary to be familiar with these equations to grasp any of the points made in this book. Indeed in most of the cases we shall consider, an analysis using the Lorentz transformation is far more complicated and less intelligible than a direct application of Rules 1 through 5. However these equations are so widely used in relativistic analysis that one would receive a rather distorted picture of the subject without them. This will therefore not be a chapter about special relativity, but about the *language* of special relativity.*

> * These remarks should not be construed to mean that the Lorentz transformation equations are useless or an unnecessary frill upon the basic facts of relativity. The point is rather that, like most abstract and concise formulations of natural laws, they come into their own primarily when one is discussing rather complicated situations in a quantitative manner. Then they are essential. However in the very simple kinds of questions we are considering here, the more direct approach is usually enough to provide explanations. Solving a problem by applying Rules 1 through 5 as opposed to solving it by applying the Lorentz transformation is rather like using simple plane geometry as opposed to analytic geometry in solving purely spatial problems. The former approach is usually far more intuitive and simple, when it works, but the latter approach is better for more complicated problems, as well as frequently supplying an almost mechanical method of solving problems even when one's intuition does not fully grasp the situation.

The Lorentz transformation tells how to relate the space and time measurements of any two inertial observers moving with velocity v with respect to each other. To derive the transformation, we first consider only events occurring along the line of their relative motion. (This specialization is only to keep the discussion a little shorter. The generalization to arbitrary events is quite simple, but it is easier to do things in two stages.)

We give each observer a coordinate system with which to describe the spatial and temporal location of events. A coordinate system consists of a long straight rod with clocks distributed along it.* The rod (called the x-axis) has marks along

> * Precisely the sort of thing pictured in the figures of Chap. 12.

it spaced 1 meter apart (as measured by a meter stick that is stationary with respect to the rod). Between each pair of adjacent marks are 100 smaller marks, each 1 centimeter apart (as measured by a meter stick stationary with respect to the rod), and between each of these, 10 millimeter marks, etc. In other words, the rod is just a long series of meter sticks laid end to end, with as many subdivisions as are necessary for the experimental accuracy one wishes to attain in making measurements of position. At each point of the rod is a clock, and the clocks are all synchronized in the rest frame of the rod in the usual manner.*

> * We assume that if clock 1 is synchronized with clock 2 and clock 2 is synchronized with clock 3, all in the rest frame of the rod, as a result of these two synchronizations clock 1 will be found to be synchronized with clock 3 in the rest frame of the rod. This can be proved for any correct method of synchronization. It is perhaps most evident if we use the method described in the Appendix to Chap. 7 of synchronizing them by direct comparison in a single place, and then carrying them very slowly to their separate stations along the rod.

With such a coordinate system, we can measure the spatial and temporal location of any event occurring along the rod by noting the mark on the rod at the point where the event occurs and the reading of the clock located at that point at the moment the event occurs. Thus I shall say that an event occurs at 100 seconds and 300 meters if it occurs right on top of the 300-meter mark as the clock there reads 100 seconds. This may seem like excessive belaboring of the obvious, but it is important to know precisely what is meant by the assertion that an event occurs at x (in the above example $x = 300$ meters) and t (in the above example $t = 100$ seconds).

Now consider two such coordinate systems moving with respect to each other along the common direction of their x-axes with velocity v. Let us arbitrarily consider one rod to be stationary, with the other sliding along it with speed v. Suppose things are so arranged that when the $x = 0$ points of the rods are adjacent, the clocks at these points read the same time, and let us call that time zero seconds.*

> * This assumption is not restrictive—it is simply a convention on where each observer shall consider the origin of his

coordinate system to be. We adopt this particular convention to make the resulting formulas as simple as possible. If, for instance, we had chosen a case in which when the two $x = 0$ points coincided, the moving clock there read 10 seconds and the stationary clock read 3, in all the Lorentz trans- formation formulas we would have to replace t' by $t' - 10$ seconds, and t by $t - 3$ seconds.

We first compare readings on the moving and stationary coordinate systems, when $t = 0$ according to the stationary one.* We used primed variables (x', t') to refer to readings

> * I have to add "according to the stationary one" because, of course, although the clock at the origin of the moving system reads 0 when the clock at the origin of the stationary one reads 0, no other moving clock reads 0 at the moment it is opposite a stationary clock reading 0. This is evident from Fig. 13.2.

of the moving system and unprimed variables (x, t) to refer to readings of the stationary one. We first ask what mark of the moving system at $t = 0$ (stationary time) is on top of the mark $x = x_0$ of the stationary system. The answer is (compare Fig. 13.1)

$$\frac{x_0}{\sqrt{1 - v^2/c^2}}.$$

$$(13.1)$$

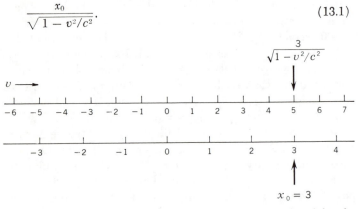

Fig. 13.1 The upper coordinate system moves to the right. At $t = 0$ (stationary time), one finds, above the mark x_0 of the sta- tionary system, the mark $x_0/\sqrt{1 - v^2/c^2}$. This is illustrated for $x_0 = 3$ in the case $v/c = \frac{4}{5}$, and hence $\sqrt{1 - v^2/c^2} = \frac{3}{5}$.

This is because the moving rod shrinks by a factor $\sqrt{1 - v^2/c^2}$ and consequently has *more* divisions between its zero point and x_0 than the stationary rod has.

Next we ask what the moving clock at x_0 reads when $t = 0$. Rule 4 tells us that it must lag behind the moving clock at 0 by an amount v/c^2 times its proper distance (13.1) from the clock at 0, or (compare Fig. 13.2)

$$\frac{- x_0 v/c^2}{\sqrt{1 - v^2/c^2}} . \tag{13.2}$$

Since the moving clock at $x = 0$ reads 0 at $t = 0$ (we are applying Rule 4 from the point of view of the stationary frame, and therefore the moving clock at x_0 is to be compared with the moving clock at 0 at the same stationary time), Eq. (13.2) must be the actual reading of the moving clock at x_0 at $t = 0$.

Having established what things are like when $t = 0$ (stationary time), we can go on to the general case. Suppose something happens at the space-time point x, t according to the stationary system. What coordinates will this same event have in the moving system? In other words, what point x' of the moving rod is on top of the point x of the stationary rod when the stationary clock there reads t, and what time t' does the moving clock at x' read at the moment of coincidence? The first half of the question is easily answered. Since the moving rod has a velocity v and the event occurs at a time t, the point of the moving rod that will be on top of the event at x at time t is just the point that was at $x - vt$ at time $t = 0$. But from

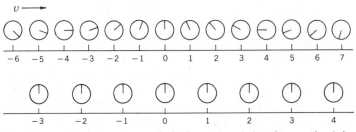

Fig. 13.2   The disposition of clocks at $t = 0$ (stationary time) in the moving (upper) and stationary (lower) coordinate systems.

(13.1) that point is just $(x - vt)/\sqrt{1 - v^2/c^2}$. Therefore

$$x' = \frac{x - vt}{\sqrt{1 - v^2/c^2}}. \tag{13.3}$$

Now the moving clock that is to be at mark x at time t will also have been at the point $x - vt$ at time $t = 0$, and therefore at time $t = 0$ it must have read

$$\frac{-(x - vt)v/c^2}{\sqrt{1 - v^2/c^2}}, \tag{13.4}$$

according to (13.2). Since then a time t has elapsed, but since the clock has moved with velocity v all that time, it has advanced by an amount

$$t\sqrt{1 - \frac{v^2}{c^2}}. \tag{13.5}$$

Therefore at time t its reading will have advanced from the value (13.4) by the amount (13.5); thus

$$t' = \frac{-(x - vt)v/c^2}{\sqrt{1 - v^2/c^2}} + t\sqrt{1 - \frac{v^2}{c^2}}. \tag{13.6}$$

With a little algebra (13.6) can be reduced to

$$t' = \frac{t - vx/c^2}{\sqrt{1 - v^2/c^2}}. \tag{13.7}$$

Thus an event that occurs at x, t in the stationary system has coordinates x', t' in the moving system, where x' and t' are given in terms of x and t by Eqs. (13.3) and (13.7). These are the two most important Lorentz transformation equations.

There are several comments to make:

1. Our analysis has been done in an asymmetrical way, since we have always considered the unprimed system to be stationary and the primed one to be moving. Now we know from the principle of relativity that we must reach the same conclusions if we consider instead the primed system to be stationary and the unprimed system to be moving with a velocity $-v$. If we had looked at it that way, all the arguments would have been the same except that primed and unprimed quantities would have been interchanged, and v would have been replaced by $-v$; hence instead of (13.3) and (13.7) we would have deduced:

$$x = \frac{x' + vt'}{\sqrt{1 - v^2/c^2}}, \tag{13.8}$$

$$t = \frac{t' + vx'/c^2}{\sqrt{1 - v^2/c^2}}. \tag{13.9}$$

In other words, (13.3), (13.7), and the principle of relativity imply (13.8) and (13.9). But from another point of view, (13.3) and (13.7) can be regarded as a pair of equations relating quantities x and t to other quantities x' and t', which can be solved algebraically to give x and t as explicit functions of x' and t'. You should check that the solution to this pair of equations is indeed given by (13.8) and (13.9), thus demonstrating that (13.3) and (13.7) are automatically consistent with the principle of relativity. This is hardly surprising, since the principle of relativity was used in constructing all the relativistic effects summarized by the Lorentz transformation; it is nevertheless gratifying to verify it explicitly, as a check that nothing has gone wrong with our analysis.

2. To generalize the analysis to the case in which events that do not occur on the x-axis are allowed for, we must consider a complete three-dimensional coordinate system, with y- and z-axes perpendicular to each other and both perpendicular to the x-axis (Fig. 13.3). We must also distribute clocks throughout three-dimensional space, all stationary and synchronized with each other in the rest frame of the three axes. The general Lorentz transformation can then be stated as follows:

Suppose a clock-axes system moves with velocity v past a stationary one in such a way that the x'-axis slides along the x-axis (as before). Let the point of space-time assigned the values $x' = 0$, $y' = 0$, $z' = 0$, $t' = 0$ in the moving system occur at $x = 0$, $y = 0$, $z = 0$, $t = 0$ in the stationary system (i.e., the origins coincide as before), and at $t = 0$ let the y'-axis fall on top of the y-axis and the z'-axis fall on top of the z-axis (which is just a convention to allow us to agree upon the relative directions of the two sets of y- and z-axes). When t is not 0,

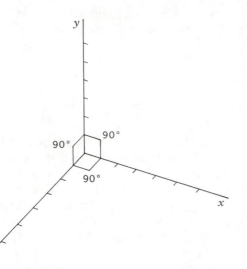

Fig. 13.3 A 3-dimensional coordinate system.

let the y and z axes remain parallel to the y' and z' axes (see Fig. 13.4). Then an event that occurs at the space-time point x, y, z, t of the stationary system has coordinates x', y', z', t' in the moving system, where

$$x' = \frac{x - vt}{\sqrt{1 - v^2/c^2}}, \tag{13.10}$$

$$y' = y, \tag{13.11}$$

$$z' = z, \tag{13.12}$$

$$t' = \frac{t - vx/c^2}{\sqrt{1 - v^2/c^2}}. \tag{13.13}$$

These results can be derived by minor generalizations of the kinds of arguments leading to (13.3) and (13.7). Rule 1 takes care of (13.11) and (13.12) since in the absence of a contraction of lengths perpendicular to the direction of motion, both coordinate systems must agree on y and z measurements. Rule 5 assures us that (13.13)—which is identical to (13.7)—and (13.10)—which is identical to (13.3)—remain valid for events lying off the x-axis, for all the moving clocks on any straight

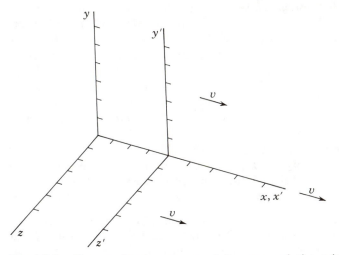

Fig. 13.4 The unprimed axes are stationary, and the primed axes move to the right with speed v. The x- and x'-axes coincide. (For clarity the scale on the x'-axis is not shown.)

line perpendicular to the direction of motion are synchronized for both observers. Thus the time that any moving clock shows is the same as the time shown by the moving clock on the x-axis directly below it, and furthermore both the primed and unprimed coordinate systems agree on which moving clock *is* directly below it (the notion of simultaneously implicit in this statement being one they both agree on). (Alternatively, a moving observer can make measurements of events lying off the x-axis by using a copy of the x'-axis rod-clock system, that is parallel to the x'-axis but displaced so the event in question lies on it. But because the stationary observer knows that clocks moving perpendicular to the line joining them and synchronized in their proper frame remain synchronized, he will not have to make any further corrections beyond those in (13.3) and (13.7).)

3. With the Lorentz transformation formulas (13.10) to (13.13), it is possible to answer all questions involving length and time measurements in a straightforward, mechanical manner, without having to devise further thought experiments. Any such question can be put in the form: If a stationary observer finds that such and such happened, what will a moving observer say? Now the stationary observer's assertion that such and such happened can be analyzed into a statement about many events, E_1, E_2, E_3, . . . all of which together make up such and such. Thus the statement "such and such happened" is really a very complicated set of statements: E_1 occurred at x_1, y_1, z_1, t_1; E_2 occurred at x_2, y_2, z_2, t_2, etc. Using the Lorentz transformation equations one can immediately conclude that the moving observer will say that E_1 occurred at x'_1, y'_1, z'_1, t'_1; E_2 at x'_2, y'_2, z'_2, t'_2, etc., where each set of primed quantities can be calculated from the corresponding set of unprimed quantities through (13.10) to (13.13).* One can then translate

> * Note that this is the principle of the invariance of coincidence again. The moving observer will agree that E_1 coincides with the point x_1, y_1, z_1 at time t_1 of the stationary system. But x_1', y_1', z_1' are just the points of the moving system coinciding with x_1, y_1, z_1, and t_1' is the time at that coincidence.

this new set of complicated statements about the many events composing such and such into what the moving observer's version of such and such will be.

This procedure is very convenient, since it enables one to stop thinking and simply perform a lot of algebraic and geometrical calculations. However it often does not make clear *why* a particular result turns out to be what it is, even though the underlying reason may be quite simple. We shall therefore avoid the Lorentz transformation whenever considerably simpler, direct arguments are available.

14

A SIMPLE WAY TO GO
FASTER THAN LIGHT
THAT DOES NOT WORK

All the formulas we have derived are valid only for objects moving with speeds less than c. In this chapter and the next we shall see that this is not a shortcoming of special relativity, since it is impossible for any material object to move faster than light.*

> * Noncolloquial version: The speed of any material object as measured by any inertial observer cannot exceed c.

This conclusion can be as jarring to the intuition as the principle of the constancy of the velocity of light. We suspect, for example, that if a rocket B moving from A at $0.9c$ should fire a second rocket C straight ahead and moving from B at $0.9c$, then the second rocket would move away from A at $0.9c + 0.9c = 1.8c$ (Fig. 14.1). This is merely a generalization of the terrestrial observation that a passenger, walking at 4 miles per hour toward the front of a train moving at 8 miles per hour through a station, moves himself at $4 + 8 = 12$ miles per hour through the station (Fig. 14.2). Although the intuitive ap-

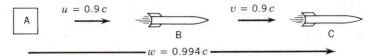

$u = 0.9\,c$ $v = 0.9\,c$

Fig. 14.1 A launches B at a speed 0.9c. Subsequently B launches C at an additional speed 0.9c. Non-relativistically one expects that C leaves A at a speed 1.8c. But the correct answer is 0.994c.

proach is right at low speeds, it is quite wrong in the case of the two rockets, where, for the particular speeds of B past A and C past B mentioned above, the correct answer is not 1.8c but about 0.994c, a speed, you will note, that is still less than that of light.

The purpose of this chapter is to show that by this apparently simple method of rockets shooting rockets shooting rockets . . . , no matter how close to c each rocket moves away from its predecessor, the final rocket will still move away from the original one at a speed less than c. We consider the general case in which B moves with velocity u with respect to A, while C moves with velocity v with respect to B.* We shall

> * By "B moves with velocity v with respect to A" we mean precisely "The velocity of B in A's proper frame is v."

only study the case in which A, B, and C lie on the same straight line; i.e., the case in which u and v are in the same

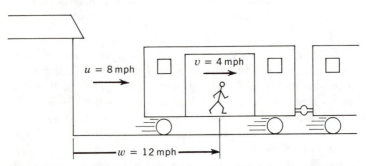

Fig. 14.2 A situation in which the non-relativistic answer is correct to a high degree of precision.

direction.* We wish to calculate the velocity w of C with

* The general case of arbitrary directions between u and v is more complicated to analyze and no more instructive; it is done in Chap. 18 (see Eq. 18.16).

respect to A.

The non-relativistic answer is simply $w = u + v$. This is certainly wrong in the case in which $v = c$, since w will then also be c: If B says that C moves with the speed of light, so will A, regardless of the relative speed of A and B. This casts some doubt on the validity of the non-relativistic formula when either u or v are close to c. The non-relativistic formula is, in fact, never exactly right. The correct result is that

$$w = \frac{u + v}{1 + uv/c^2}. \qquad (14.1)$$

This is where the 0.994 comes from: $0.994 \approx \dfrac{0.9 + 0.9}{1 + (0.9)^2}$.

There are many ways to derive (14.1). The simplest analytically is to consider C to be equipped with a flashing clock that takes a time t between ticks in its rest frame. Then B, watching C's clock, will see a flash every $f_v t$ seconds (B-seconds), where f_v is the slowing down factor introduced in Chap. 5, given by*

$$f_v = \sqrt{\frac{1 + v/c}{1 - v/c}}.$$

* See the second equation in the Appendix to Chap. 6.

These flashes go past B and are observed by A, who can regard them as coming either from C's clock, with its proper time t between ticks or, equally well, as coming from a flasher of B's which has a proper time $f_v t$ between ticks.* From the

* For B could make such a flasher and adjust it to flash each time he received a flash from C; the flashes from C and those from B's flasher would then arrive together at A, since they would leave B at the same moment.

former point of view, A will see a flash every $f_w t$ seconds, but from the latter point of view he must see a flash every $f_u(f_v t)$

seconds. Since both points of view are permissible, they must lead to the same conclusion, and therefore *

$$f_w = f_u f_v. \tag{14.2}$$

* Notice that the result $f_v f_{-v} = 1$ derived in the Appendix to Chap. 5 is just a special case of 14.2), since if $u = -v$, $w = 0$.

If we substitute into (14.2) the explicit form the f's have we find

$$\sqrt{\frac{1 + w/c}{1 - w/c}} = \sqrt{\frac{1 + u/c}{1 - u/c}} \sqrt{\frac{1 + v/c}{1 - v/c}}. \tag{14.3}$$

Squaring both sides of (14.3) gives

$$\frac{1 + w/c}{1 - w/c} = \frac{1 + u/c}{1 - u/c} \frac{1 + v/c}{1 - v/c}, \tag{14.4}$$

which gives w in terms of u and v. It is not hard to solve this for the unknown w and verify that the solution is just (14.1).

Equation (14.1) is sometimes known as the relativistic addition law for parallel velocities, a somewhat misleading name, since only in the non-relativistic case is w simply the sum of u and v.

When u and v are both much less than c, then $1 + uv/c^2$ is very close to 1, and (14.1) differs very little from the old law, $w = u + v$. However when either u or v is equal to c, then (14.1) requires that w also be equal to c, giving us back the principle of the constancy of the velocity of light. At speeds between c and non-relativistic ones, Eq. (14.1) expresses the impossibility of attaining speeds greater than c in stages, for if u and v are both less than c, w must also be less than c.*

* Proof: It is enough to show that $x + y$ is always less than $1 + xy$ if x and y are less than 1. (Here x and y are u/c and v/c.) This in turn will be true if $1 + xy - x - y$ is positive, but this last expression is just $(1 - x)(1 - y)$, and since x and y are less than 1, each factor in the product is positive.

It is therefore impossible to get an object moving faster than light by a succession of pushes, each of which increases its velocity by any amount less than the velocity of light. If an

object is to move faster than light, it cannot attain the speed
gradually, but must do so, in some sense, "all at once," whatever that may mean.* This consideration makes it seem unlikely

* As it turns out it means nothing at all, since nothing can,
in fact, go faster than light.

that anything can go faster than light but does not rule out the
possibility that something might be moving with a speed
greater than c that it had attained in some unknown way (or
that it always possessed). This possibility can also be ruled out,
however, on rather different grounds, which we take up in the
next chapter.

It is a remarkable fact that an experiment verifying the relativistic addition law was performed in the mid-nineteenth century, decades before the development of special relativity. The
experiment measured the speed of light in moving water.
Water flowed through a pipe with speed v, and light was trans-
mitted through the moving water from one end of the pipe to
the other. We cannot understand this experiment by a simple
direct appeal to the constancy of the velocity of light, since it
is only the speed of light in vacuum that is the same for all
observers. The speed of light in water is only about $\frac{3}{4}c$ with
respect to the water (since the velocity is no longer c, we must
add with respect to what it is being measured) and therefore
if we wish to analyze the experiment by looking at it in different inertial frames, we shall have to use (14.1) to transform
the velocities. Conventionally the speed of light in a trans-
parent medium is written as c/n, where n is called the index
of refraction of the medium. The index of refraction of water is
about 1.33. The result of the experiment was that the speed w
of light in moving water was its speed c/n in stationary water
plus a correction proportional to but less than the speed of the
water:

$$w = \frac{c}{n} + v \left(1 - \frac{1}{n^2}\right). \tag{14.5}$$

Non-relativistically this is very hard to understand. In the
rest frame of the water the speed of the light is c/n and
since the water moves with speed v in the direction of the

light, the net speed according to the old addition law would be $c/n + v$. There are further complications and possible ways out of the dilemma if one is so non-relativistic as to believe in an ether, for then the speed of the light in the moving water would depend on the speed of the water with respect to the pipe *and* on the effect of the moving water on the ether. People did in fact produce ether theoretic explanations of (14.5), but none were as simple as the relativistic explanation:

The speed of the light with respect to the water is c/n, and the speed of the water with respect to the pipe is v; therefore the speed w of the light with respect to the pipe is just (from (14.1))

$$w = \frac{c/n + v}{1 + v(c/n)/c^2}.$$

An algebraically equivalent way of writing this formula is

$$w = \frac{c}{n} + v\left(1 - \frac{1}{n^2}\right)\frac{1}{1 + v/nc}.$$

This differs from the experimentally determined formula (14.5) only in an extra factor $1/(1 + v/nc)$ multiplying the coefficient of v. Since the speed of the water is enormously smaller than c, this factor is almost exactly 1, and its detection is well beyond the accuracy of the experiment.

15

WHY NOTHING CAN GO FASTER THAN LIGHT

The usual way of achieving high velocities, by continually increasing the velocity by small amounts, can never result in a speed greater than light. If (in a given frame) the speed of an object is less than that of light, then if its speed is increased (in its proper frame) by any amount less than that of light, the final speed of the object (in the original frame) will still be less than that of light.* This makes it difficult to understand

* Note that although the velocity of an object depends on the frame in which it is measured, if a given inertial observer finds that a particular object moves faster than (or slower than or at the same speed as) light, any other inertial observer will reach the same conclusion. For the object can be made to run a race with a photon (cf. the discussion at the end of Chap. 2) the speed of which is, of course, c for all inertial observers. The outcome of the race is something all observers must agree on, for different effects can be produced at the finish line depending on whether the object or the photon gets there first or whether the race is a draw.

how speeds greater than c could be attained, but does not exclude the possibility that objects could travel faster than light.

There is, however, a simple direct reason why no material object can travel faster than light, quite independent of questions relating to how the velocity was actually attained. For suppose a ship traveled from the Earth to Sirius at a speed u greater than c. Let us describe its journey in a coordinate system in which the Earth is at rest at $x = 0$ and in which the x-axis is along the line from Earth to Sirius, which is a distance l from the Earth. Suppose the ship leaves the Earth at $t = 0$ and arrives on Sirius at $t = T$. Since its speed is u in the Earth's frame, $T = l/u$. Thus in the Earth's frame, the departure of the ship from Earth takes place at $x = 0$, $t = 0$, and its arrival at Sirius is at $x = l$, $t = l/u$.

Consider, now, a second observer moving along the line from Earth to Sirius at a speed v. Let us give him a coordinate system whose x'-axis lies along the Earth-Sirius line (and hence along the x-axis), and let him also assign the coordinates $x' = 0$, $t' = 0$ to the departure of the ship from Earth. Then the coordinates he assigns to the arrival of the ship at Sirius are given by the Lorentz transformation equations (13.10) and (13.13), where x is taken as l and t as l/u. Hence

$$x' = \frac{l - v(l/u)}{\sqrt{1 - v^2/c^2}} = \frac{l(1 - v/u)}{\sqrt{1 - v^2/c^2}}, \tag{15.1}$$

$$t' = \frac{l/u - vl/c^2}{\sqrt{1 - v^2/c^2}} = \frac{l/c^2}{\sqrt{1 - v^2/c^2}} \left(\frac{c^2}{u} - v \right). \tag{15.2}$$

In particular, the moving observer finds that the ship departed at $t' = 0$ and arrived at t' given by (15.2). If the arrival is to come after the departure, t' had better be positive. This will be so provided $c^2/u - v$ is positive. Now if u, the speed of the ship in the Earth's frame, is less than c, then c^2/u is greater than c, and $c^2/u - v$ is indeed positive for any v less than c. On the other hand if u exceeds c, then c^2/u is less than c, and we can choose a velocity v for the moving observer which is less than c and yet so large that $c^2/u - v$ is negative. For such an observer, the ship arrives at Sirius *before* it leaves the Earth. The only way such an absurdity can be avoided is if nothing can go faster than light.

The same conclusion follows directly from Rule 4. The observer moving from Earth to Sirius at speed v notes that the Earth clocks read 0 when the ship leaves the Earth, and the Sirius clocks* read l/u when the ship arrives. Now for the

* Also stationary in the Earth's frame, the relative velocity of Sirius and the Earth being negligible.

moving observer the Earth-Sirius system moves parallel to its length with speed v, and hence the clocks in front (Earth clocks) lag behind the clocks in the rear (Sirius clocks) by lv/c^2. Thus at the moment the Earth clocks read 0, the Sirius clocks read not 0, but lv/c^2. If v exceeds c^2/u, this is greater than their reading l/u when the ship arrived on Sirius. Hence the moving observer, since he knows that the clocks on Sirius are ordinary clocks that do not run backward, must conclude that the ship arrived on Sirius before it left the Earth.

A proper relativist might argue at this point that the controversy is purely verbal and that any observer who says the arrival happened before the departure would simply reinterpret things, calling the arrival the departure and vice versa. Thus some observers would say that the ship moved from Earth to Sirius, while others would say that it moved from Sirius to the Earth. But this is a difference of opinion which, unlike some we have encountered earlier, we cannot tolerate. The material presence of an object can cause permanent effects, which enable us to distinguish the temporal sequence of events. Suppose, for example, the ship crash-landed near Sirius. Then it either would or would not be scratched and dented when it left Earth, depending on whether the departure from Earth was after or before the landing near Sirius. Simply by inspecting the ship carefully on Earth, we could determine whether the crash near Sirius had happened or had not yet occurred.*

* If one were stubborn, one could adopt the point of view of the frame in which the crash near Sirius occurred first, even though the ship were unsullied while on Earth, by arguing that somehow the scratches were undone between the Sirius

crash and the Earth takeoff. In fact the entire progression from Sirius to Earth would appear rather like a movie in reverse. A cloud of dust would converge on a dented ship. The ship would, while losing its bruises, rise from Sirius by a process of converting exhaust gases back into liquid fuel and in this manner arrive back on Earth, unscratched and with a full fuel tank. If you pursue this line still further, you will discover that while the ship is taking this mysterious journey from Sirius to Earth, it is also in two other places at the same time! It is sitting shiny and new, with a full tank, on its launching pad on Earth, and it is lying wrecked near Sirius. You will also notice two curious events: When the ship leaves Sirius for Earth it leaves a copy of its bruised self behind; somewhat later, when it arrives on Earth, it merges with its intact self, immediately after which, both disappear. Such goings-on are sufficiently unnerving and inconsistent with laws of mechanics and thermodynamics, which must hold for any inertial observer, that the only acceptable alternative is to rule out the possibility of anything moving faster than light.

The general reason excluding travel faster than light is that if an object could move that fast, there would be inertial observers who disagreed on the time order of two events one of which was capable of materially influencing the other through the agency of the rapidly moving object. The relation between cause and effect could then be reversed simply by examining the same events in different inertial frames. In any concrete case this leads to absurdities like the ones considered above, forcing us to deny the possibility of moving anything faster than light. Because this denial is forced upon us by our commitment to the preservation of causal relationships, the fact that no object can move faster than light is sometimes known as the principle of causality.

As another way of looking at the impossibility of speeds in excess of c, let us reconsider the non-relativistic apparatus of Chap. 12. Suppose that while the two pictures were being taken, a candle had moved along stick A at a speed of 8 meters per second* (Fig. 15.1), passing mark 5 on stick A when the

* Remember that the speed 4 meters per second played the role of c in this model.

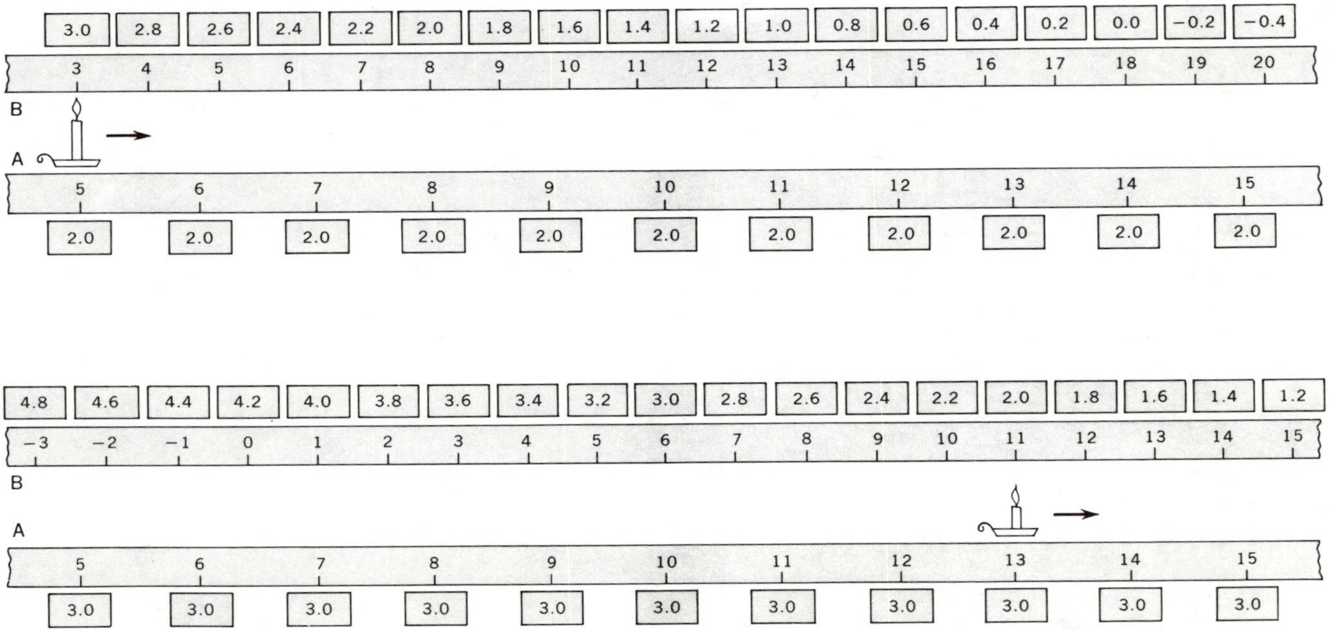

Fig. 15.1 Figs. 12.1 and 12.2, their mystery betrayed by a moving candle which proves conclusively that system A is the correct one. (Provided, of course, we know the two pictures are of the same candle.)

clock there read 2 and mark 13 on stick A when the clock there read 3. Since the candle burns along the way, it is shorter at mark 13 than at mark 5.

Now according to system B, the true time at which the candle passed A's mark 5 was 3.0 seconds, the reading of stick B's clock directly above it, and the true time at which the candle passed A's mark 13 was 2.0 seconds. Hence according to system B the candle was as seen in the lower half of Fig. 15.1 1 second before its appearance in the upper half. This is impossible (candles do not get longer as they burn), and therefore system A must be the correct one.

In other words, objects moving at speeds greater than 4 meters per second in the pictures of Chap. 12 could give us a way of telling which system was the correct one. Now in the real world, c is 2.9979 hundred thousand kilometers per second, and the principle of relativity tells us that it is impossible to decide which inertial system is "correct." Therefore there can be no objects in the real world which travel faster than 2.9979 hundred thousand kilometers per second.

16

THE CLOCK "PARADOX"

The fact that A and B each finds that the other's clocks run slower than his, is paradoxical only because one is used to thinking of the rate of a clock as something inherent in the clock, independent of how it is measured. If one looks at things more carefully, one sees that the assertion that a clock runs at a certain rate is actually a compact way of summarizing a series of observations and deductions. Since the observations and deductions A makes to ascertain the rate of his and B's clocks are different from those made by B, there is no contradiction in the finding of each that the other's clocks run slowly. Any remaining sense of contradiction one is left with should be due only to the strength of irrational linguistic habits.

Nevertheless one can still get confused, and it is instructive to examine cases in which one tries to force the relativistic effects into producing an impossible conclusion, to see how nature escapes from the difficulties.

The most celebrated confrontation of relativistic reality and confusion is the so-called clock paradox. Suppose a spaceman

B goes on a high-speed rocket journey, then turns around and returns to Earth. According to an observer A on the Earth, B's clocks run slowly throughout his trip; hence when the space clocks are compared with the Earth clocks upon their return, they will be found to have advanced less. Indeed, B himself must age less than A does during the journey, since all processes, biological, chemical, or otherwise, taking place within B must go slowly to the same extent his clocks do. Otherwise B could distinguish between a state of rest and one of uniform motion by comparing the rate of his clocks with the rate he himself was functioning at. Thus if A and B were twin brothers, A would be physically older than B at the end of B's journey. (For this reason the clock paradox is also known as the twin paradox.)

To the non-relativist, all this is perhaps surprising, but not paradoxical. The apparent paradox emerges when it is suggested that since B can consider himself to be at rest and A to be moving, B can conclude that A's clocks run slower than his and will therefore expect A to be younger than himself when the two meet again. Which one is right?

The answer is that when B gets back, he will indeed be younger than A. The fallacy in the reasoning leading to the opposite conclusion is in the words "B can consider himself to be at rest and A to be moving." This is not true.* It appears to

* At least it is not true within the framework of special relativity; further aspects of the "paradox" are qualitatively considered from the point of view of general relativity at the end of this chapter, but it should be realized that there is no need to appeal to general relativity to show that there is no paradox.

follow from the principle of relativity only when we forget that B is *not* an inertial observer. Only when two observers move *uniformly* with respect to each other is each entitled to regard himself as at rest. In the present case B does not move uniformly with respect to the inertial observer A, so although A is in an inertial frame, B is not.

This lack of uniformity in B's motion is best illustrated by plotting B's position against time as seen from A's point of view

(Fig. 16.1). The curve representing anything moving uniformly with respect to A would be a straight line. B's curve, however, is a straight line *with a bend in it* at the moment he turns around. Of course B *is* moving uniformly with respect to A at all times except for the brief moment when he turns around, but this is quite enough to make the total curve in Fig. 16.1 very different from a single straight line.

The nonuniformity of B's motion with respect to A is not simply an abstract observation about graphs, but results in definite physical effects that destroy the superficial symmetry between B and A. To turn his ship back toward Earth B must fire braking rockets and then fire rockets until he has built up a velocity in the opposite direction. During the period of deceleration he feels forces of many g's, and the occurrence of such forces at this point in his journey makes it clear to him that he is not in an inertial frame throughout the trip. A, however, fires no engines, feels no forces, and is in an inertial frame from the moment of B's departure to his return.

Thus the analysis from A's point of view, concluding that B is the younger at the end of the journey, is correct, since A

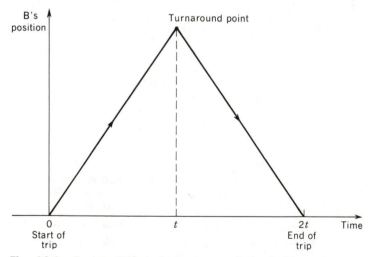

Fig. 16.1 A plot of B's trajectory as made by A. The trajectory of an inertial observer would be a single straight line.

is in an inertial frame and entitled to use Rules 1 through 5. The analysis leading to the conclusion that A is the younger is fallacious, since it is based on the incorrect assumption that B is an inertial observer.

But although the simple observation that B is not in a state of uniform motion eliminates the "paradox," there are still several possibilities for confusion in analyzing in detail why both A and B must agree that B should be younger.

To begin with, although it is the forces of deceleration that distinguish B from A and reveal to B that he is not an inertial observer, one would nevertheless have a very misleading view of the situation if one thought that B's lesser aging was a direct result of the action of these forces. For consider a second journey of B's, twice as long as the first. Since his clocks run slowly for twice as long, he will find on returning that he has aged less than A by twice the amount that he did on his first journey. However the process by which he turned himself around can be exactly the same as it was the first time. Therefore it will have the same effect on his aging. Whatever this effect is, we can make it as small as we wish compared with the total difference in aging simply by considering a sufficiently long journey. We need therefore not worry about what happens while B turns around as a possible explanation for the difference in aging between A and B; it cannot have anything to do with it.

The only important thing about B's turning around is that it means that B is not in a single inertial frame throughout the journey, but in two different frames* (one on the way out, and

---

* Plus a host of different ones while he turns around, which we have just shown are not of any great importance.

---

one on the way back). Thus we can describe the trip from B's point of view *provided* we remember that B changes coordinate systems halfway through his trip; i.e., to be able to describe the journey from the points of view of either A or B, we need to consider not two inertial observers, but three: one stationary with A, one moving uniformly with respect to A and stationary with respect to B as B goes out, and one moving uniformly with respect to A and stationary with respect to B as B comes back in.

We can continue to call the first of these observers A, but we now need two observers for B, B_1 and B_2. B_1 moves uniformly with B as B makes the outward journey (and continues moving uniformly outward after B turns around), and B_2 moves uniformly with B as B makes the return journey (and had been moving uniformly toward A with the same velocity before B turned around). The positions of B_1 and B_2 are plotted against time from A's point of view in Fig. 16.2. Each is described by a straight line, since each is moving uniformly with respect to A. Comparing Figs. 1 and 2 one sees that B moves along B_1's trajectory between times 0 and t and along B_2's between times t and $2t$.

To make things concrete, suppose the journey is to a point somewhere in the vicinity of Vega, exactly 25 light-years from the Earth, at a speed of $0.995c$, for which $\sqrt{1 - v^2/c^2} = 0.1$.

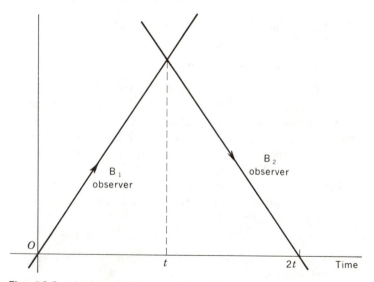

Fig. 16.2 A plot of the straight line trajectories of two inertial observers B_1 and B_2, as made by A.

Let the B_1 observers be passengers on an express rocket that passes with uniform speed v past Earth and then past Vega.*

> * Passengers must jump on and off, since the express, being an inertial frame, does not stop. The boarding process may be hard on the passengers, but this need not concern us, since the manner in which the accelerations of B are performed has no bearing on the resolution of the paradox.

The B_2 observers live on a return shuttle, passing uniformly with speed v from Vega to Earth and beyond. The timetable is such that the two shuttles meet at Vega, so that B does not have to wait before transferring.

From the point of view of those on Earth, B takes a time $l/v = 25$ years $1\frac{1}{2}$ months (l is 25 light-years) going out, and another l/v coming back, so that $2l/v$, or 50 years 3 months elapse on Earth between B's departure and his return. However throughout his journey* B's clocks have been running

> * Except, perhaps, for the brief moment of transfer, which, as we have mentioned earlier, cannot be important.

slowly and have therefore advanced by only $\sqrt{1 - v^2/c^2}$ $(2l/v) = 5$ years.*

> * We shall round off all times to the nearest half month.

Thus B, returning from what he experienced as a 5-year journey, finds the friends he left behind old enough to be his grandparents. How can he explain this in view of the fact that everyone in the outgoing shuttle (which *is* an inertial frame) agreed that the Earth clocks were running *slower* than B's, as did everyone in the incoming shuttle?

From B's point of view it looked like this:

On the outgoing shuttle Earth receded and Vega approached at speed $v = 0.995c$. The distance between Earth and Vega was only $l\sqrt{1 - v^2/c^2} = 2\frac{1}{2}$ light-years,* and therefore it took

> * We shall also round off all distances to the nearest half light-month.

$(l\sqrt{1 - v^2/c^2})/v = 2$ years 6 months for the Vega station to arrive. Similarly, on the return shuttle Earth had to return a distance of $2\frac{1}{2}$ light-years at a speed of $0.995c$, and so the return journey also took about 2 years 6 months.

B therefore agrees with the analysts on Earth that he should be $(2l\sqrt{1-v^2/c^2})/v$, or about 5 years older when he returns. It is tempting to add that he will find the people on Earth, whose clocks he knows have been going slowly, only $[(2l\sqrt{1-v^2/c^2})/v]\sqrt{1-v^2/c^2} = 6$ months older than when he left. This, however, is wrong. To find the mistake, let us follow B step by step through his journey. When clocks in the Earth station indicate time zero, B waves goodbye to his friends, hops aboard the outgoing shuttle, and immediately asks the occupants some questions:

Q. How far is it to Vega?

A. $2\frac{1}{2}$ light-years.

Q. What does the clock on Earth say right now?

A. It says zero, since we've just picked you up and we're on schedule.

Q. What does the clock on Vega say right now?

A. Since it's in the Earth frame, synchronized with the Earth clock in that frame, and 25 light-years behind the Earth in the Earth frame, it must be (Rule 4) ahead of the Earth clock by $lv/c^2 = 24$ years $10\frac{1}{2}$ months. Since I just told you the Earth clock reads zero, the Vegan clock now reads 24 years $10\frac{1}{2}$ months.

During the $2\frac{1}{2}$ years it takes Vega to reach the outgoing shuttle, B knows that the clocks on Earth and Vega are running slowly, and advance by only one-tenth of $2\frac{1}{2}$ years, or 3 months. Thus just before B transfers he notes that the time on Earth is only 3 months after his departure, while the clock on Vega has advanced by 3 months, bringing it up to 25 years $1\frac{1}{2}$ months. As B jumps aboard the return shuttle he glances at the Vegan clock, which does indeed read 25 years $1\frac{1}{2}$ months* and again

> *Which is also the time his friends on Earth expected him to be transferring.

starts to question the passengers:

Q. How far is it to the Earth?

A. $2\frac{1}{2}$ light-years.

Q. What does the clock on Vega say right now?

A. It says 25 years $1\frac{1}{2}$ months, since we've just picked you up and we're right on schedule.

Q. What does the clock on Earth say right now?

A. Since it's in the Vegan frame, synchronized with the Vegan clock in that frame, and 25 light-years behind Vega in the Vegan frame, it must be (Rule 4) ahead of the Vegan clock by $lv/c^2 = 24$ years $10\frac{1}{2}$ months. Since I just told you the Vegan clock reads 25 years $1\frac{1}{2}$ months, the Earth clock now reads 50 years.

"Of course," muses B, "the old relativity of simultaneity correction," makes the appropriate adjustments in his travel diary, and notes during the $2\frac{1}{2}$ years of the return journey that the Earth clocks, running at only one-tenth of his rate, advance by 3 months, which brings them up to 50 years 3 months at the time of his return.

The fallacy that would have led B to expect his friends on Earth to be only 6 months older than when he left, had his fellow passengers not corrected him, is the following: Although it is true that according to the outgoing frame only 3 months elapse on Earth during B's outward voyage, and according to the incoming frame only another 3 months elapse during B's return journey, in changing ships B enters a frame in which the current time on Earth is 49 years 9 months later than it is in the frame he left. The missing 49 years 9 months is all in the correction due to the differing notions of simultaneity in frames B_1 and B_2.

If that strikes you as less than satisfactory, as involving a notation in B's travel diary rather than a real process occurring on Earth, let me point out that the original paradox was also based on statements B was making about reality, rather than the reality itself. B's assertion that 3 months had elapsed on Earth at the moment before he transferred was no less (or more) remote from reality than his subsequent 49-year 9-month correction. The farther one gets from a place, the more a matter of convention one's assertions become about the correct time "now" in that place. At the moment of B's transfer, inertial observers in the neighborhood of Vega could have been found maintaining that the time on Earth was anything between $1\frac{1}{2}$ months, and 50 years $1\frac{1}{2}$ months, after B's departure. One stays with the measurements and deductions one has made in a given frame not because they have a truth that transcends that frame, but because as long as one remains in that frame they can never lead to false conclusions. However one must be prepared to correct them when one moves to another frame. In the case of the clock paradox, the correction supplies almost all the effect.

One could probably continue to ponder this with sophistication and subtlety, but the question is closer to philosophy than to physics.* As an antidote to such speculations, we can re-

* Indeed, at this point it is not clear what the question is.

describe the events in a rather different manner, concentrating exclusively on what B observes, rather than on what he says or is told. This time we shall not allow him to talk to anyone aboard either of the ships but shall require him steadfastly to keep his telescope trained on his friends on Earth, whose aging he can thereby keep continuous track of.

To help B, his friends each wear a lightbulb that flashes once a second in its proper frame. In the outgoing frame B_1, the Earth recedes with speed v, and the bulbs flash only once every $1/\sqrt{1 - v^2/c^2} = 10$ seconds, since they move with the Earth. Between flashes, the bulbs get a distance $v/\sqrt{1 - v^2/c^2} = 10$ light-seconds* farther away, which extra distance each

* Actually closer to 9.95 light-seconds, but this is a small enough error to have no serious effect on our answer.

extra flash has to travel before it reaches the shuttle. Hence flashes arrive at the shuttle only once every 20 seconds.*

* $20 = 10$ (the time between flashes) $+ 10$ (the extra time it takes each subsequent flash to travel its extra 10 light-seconds).

As we noted earlier, B spends $2\frac{1}{2}$ years on the outbound shuttle, and since he receives flashes at one-twentieth of the rate they occur on the Earth, at the moment of transfer he has seen one-twentieth of $2\frac{1}{2}$ years, or $1\frac{1}{2}$ months of (slow motion) events on the Earth.

He changes ships swiftly at Vega, eye glued to the telescope, missing no flashes, and continues counting them in frame B_2 of the incoming shuttle. In frame B_2 the Earth still moves with velocity v, and so the bulbs still flash only once in 10 seconds. However the Earth now moves toward the shuttle, advancing 10 seconds $\times$ 0.995c between flashes, and so successive flashes now have 9.95 fewer light-seconds to travel, as a result of which a flash is received every twentieth of a second.*

> * $1/20 = 0.05 = 10$ (the time between flashes) $- 9.95$ (the time saved by each subsequent flash because of its having 9.95 fewer light-seconds to travel).

Thus on the inbound journey the fact that Earth is racing toward the ship more than compensates for the fact that the Earth flashers flash slowly because of their motion, and B sees 20 flashes per second. He therefore sees his friends aging 20 times as fast as he ages, and since he watches this for the entire $2\frac{1}{2}$ years of the return journey, he sees 50 years $1\frac{1}{2}$ months* pass on Earth. Thus B sees the entire 50 years 3

> * The extra $1\frac{1}{4}$ months needed to make the total aging on Earth equal to 50 years 3 months comes from the fact that we rounded off the time B spent coming back to an even $2\frac{1}{4}$ years. In this one instance we need the more accurate figure 2.509 years, and we also need the more accurate figure of 19.975 Earth flashes received per second. ($\sqrt{1 - (0.995)^2}$ is not precisely $1/10$.) The reason we have to be a bit more careful in this one case is that we multiply by 20 at the end, and therefore all the mistakes get multiplied by 20. This still leaves us very close to the right answer (50 years as compared with 50 1/8 years), but we must be accurate to within better than 1/8 year, if we want the final answer to be right to the nearest half month.

months that his friends on Earth age, $1\frac{1}{2}$ months of it on the outgoing trip and all the rest on the way back in. On both legs of the journey he knows that the Earth clocks are running slowly. On the way out he is moving away from the flashes, and so he sees the aging more slowly still, but on the way back he is moving toward the flashes so rapidly that the aging he sees speeds up to allow him to watch over 50 years of it in the $2\frac{1}{2}$ years of the trip back in.*

> * However I must emphasize again that B watches an aging process that is going on at a slowed up rate on both legs of his journey. He sees it sped up on the return trip only because his motion toward the flashes of light is so rapid that it more than compensates for the slow rate of flashing. To emphasize this, note that if, for some reason, the Earth flashers did not run slowly in frame B_2, B would see not 50, but 500 years of Earth time flashed off on his return trip.

We can also verify that if B's friends on Earth watch him with their telescopes, they will see him age 5 years during his trip. As B recedes with speed v, they will observe one flash every 20 seconds from B's one-second-proper-time flasher, since the flasher slows down to a flash every 10 seconds, and light from subsequent flashes takes about 10 seconds more to get back to Earth. This applies to every flash B emits on his outward journey, and since the last flash B emits while on the outgoing shuttle takes 25 years to get back to the Earth (having been emitted 25 light-years away), B's friends will see flashes at the rate of one every 20 seconds for those 25 years plus the 25 years $1\frac{1}{2}$ months B spent on the outward shuttle.

Hence during the first 50 years $1\frac{1}{2}$ months on Earth after B departs, his friends will see him age about one-twentieth of this amount, or $2\frac{1}{2}$ years. During the remaining month and a half before his return, however, his friends will be receiving the incoming flashes, which arrive every twentieth of a second, since although they are still emitted 10 seconds apart, B is 9.95 light-seconds closer for each. Hence in that last month and a half they will see B age 20 times as fast, which accounts for the other $2\frac{1}{2}$ years.

Before concluding our discussion of the clock paradox, I should mention that it is sometimes stated that the clock paradox can be resolved only within the framework of the *general* theory of relativity (as opposed to the theory we have been examining up till now, the *special* theory of relativity). This is not true, there being no paradox within the framework of the special theory. What such statements mean is that if we

wish to describe things from the point of view of B *and* to regard A and B as equivalent observers (which they are not in
the special theory, since A is in a single inertial frame while
B changes frames halfway through the trip), general relativity
must be resorted to.

The general theory of relativity allows one to describe events
from the point of view of an observer in an arbitrary non-
uniform state of motion, and we require this more powerful
form of description if we wish to allow B to regard himself as
at rest throughout the entire journey. Once B decides that he
is at rest at all times, however, he has some rather sticky points
to account for. He must explain why it is that although during
a certain short period he experienced immense forces (those
encountered while turning around, from the point of view
of Earth), nevertheless he did not move. Furthermore he must
explain why it is that the direction of motion of the Earth and
Vega suddenly reversed, although people on Earth and in the
Vega station seem to have felt no jarring forces at all. In the
general theory he can account for this by saying that at a
certain instant a gravitational field appeared, which had the
effect of changing from v to $-v$ the velocity of Earth, Vega,
and everything and everybody else, except B, who at that
moment resisted the action of the field by applying his braking
rockets (or by jumping out of his stationary ship before the
action of the field into a ship which became stationary after
the action of the field). The forces B felt were thus the result
of his efforts to counter the action of the gravitational field and
remain at rest in spite of it. The people on Earth and in the
Vega station were unaware of this sudden gravitational field
in much the same way that a person in a state of free fall does
not feel the gravitational forces acting on him.

In deciding whether he will be younger than his friends on
Earth when the journey ends, B must take account of one of
the basic facts of the general theory, that clocks in different
parts of a gravitational field run at different rates. If B does the
appropriate calculations he will indeed find that even though
he was stationary and the Earth moved throughout the journey, so that most of the time the Earth clocks were running
slower than his, nevertheless, while the gravitational field

acted, the Earth clocks sped up enough to make the total aging of his friends on Earth greater than his.*

* Our earlier argument that what happens during the time of deceleration can have no effect no longer applies. This argument used the principle that physical events occurring in one part of space will not happen any differently if only their location is altered. However once a gravitational field is present, different regions of space are no longer equivalent. If we try the same argument as earlier, asking what happens if the journey takes twice as long, then B will be twice as far from the Earth at the moment the gravitational field arises, and the difference in gravitational potential energy between B and the Earth will be twice as great. It turns out in the general theory of relativity that this is just enough to cause the speeding up of the Earth clocks due to the gravitational field, to increase by the amount necessary to give the right answer again.

17

MINKOWSKI DIAGRAMS: THE GEOMETRY OF SPACE-TIME

There is a different way of deriving and representing the relativistic effects contained in Rules 2 through 4, which uses graphs, geometry, and trigonometry rather than the combination of thought experiments and algebraic analysis we have so far relied on. The alternative method is somewhat more abstract but provides a simple, unified picture of all the relativistic effects in which their mutual self-consistency is obvious.

The diagrammatic approach, due to Minkowski, can either be presented as a way of displaying graphically the Lorentz transformation equations, or it can be derived directly from the principles of relativity and the constancy of the velocity of light, as Rules 2 through 4 were. The second approach is undoubtedly the prettier of the two, and we shall follow it here. When finished, we shall be able to rederive the Lorentz transformation from the diagrams.

We shall consider how two inertial observers, moving with velocity v relative to each other, describe events taking place along a line parallel to v. The restriction to events taking place

along a single line, i.e., to a single spatial dimension, is made
only to keep the diagrams simple. The extension to the description of events occurring in three-dimensional space involves no new concepts, and will be mentioned at the end of
the chapter.

First consider a single observer A who describes events
occurring along a line in terms of the time t at which the event
occurs and the distance x of the event from some fixed reference point. A can represent the location of an event in space
and time by a plane graph in which the vertical axis represents
t, and the horizontal, x (Fig. 17.1). The point P in Fig. 17.1, for
example, represents an event taking place 1.5 meters to the
right of the origin at a time 2.5 seconds after time zero (or,
more briefly, at 1.5 meters and 2.5 seconds).

A can also use this diagram to describe the entire history
of a point object (or particle) by indicating on the diagram
where it is at every instant of time, thereby producing a line
rather than just a point. Figure 17.2, for example, shows the

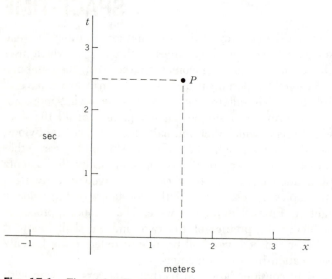

meters

Fig. 17.1 The point P represents an event that occurs
at 1.5 meters to the right at 2.5 seconds after time zero,
according to the rectangular system of coordinates.

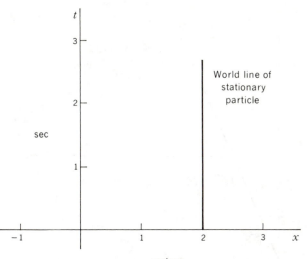

meters

Fig. 17.2   World line of a stationary particle. According to the rectangular system of coordinates the heavy line represents a particle that is stationary at 2.0 meters to the right between time zero and a little less than 3 seconds.

line corresponding to an object which is stationary at $x = 2$ meters, while Fig. 17.3 shows the line describing an object moving uniformly to the right. We can deduce the velocity of the object from the diagram, for it takes 2 seconds for the object to move each additional meter to the right, and hence its velocity is half a meter per second.

Lines describing the history of point objects are called world lines, or space-time trajectories. The world line of an object stationary with respect to A is just a vertical line, parallel to the t-axis in A's diagram; the world line of an object moving with constant velocity is a straight line *not* parallel to the t-axis; and the world line of a nonuniformly moving object will, in general, be curved, an example of which is shown in Fig. 17.4.

Since the speed of light is about 300 million meters per second, the world line of a photon is barely distinguishable from a horizontal line, its t-coordinate advancing about three-

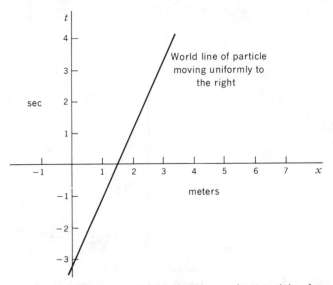

Fig. 17.3   World line of a uniformly moving particle. According to the rectangular system of coordinates the heavy line represents a particle that moves from slightly to the left of 0 meter a little more than 3 seconds before time zero, to slightly to the right of 3 meters at a little more than 4 seconds after zero.

billionths of a second for each change of 1 meter in its x-coordinate. Thus the scales of the t- and x-axes in Figs. 17.1 to 17.3 are extremely inconvenient for discussing the behavior of objects moving anywhere near the speed of light. The world lines of such particles are so close to being horizontal as to be indistinguishable from the world line of a hypothetical object with infinite velocity (which would be perfectly horizontal).

To represent velocities near c easily, we must enormously compress the x-axis. This is most conveniently done by compressing the scale on the x-axis to the point where the segment of x-axis representing the distance a photon travels in a given time interval has the same length as the segment of t-axis representing that interval. In other words we shall continue to measure the t-axis in seconds but shall measure the x-axis in light-seconds (i.e., the distance light travels in a second,

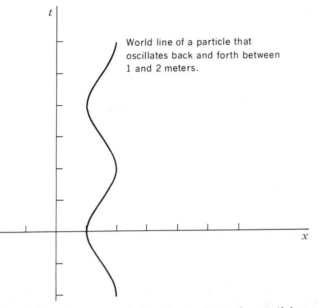

Fig. 17.4   World line of a nonuniformly moving particle.

2.9979 hundred thousand kilometers), letting a distance on the *t*-axis corresponding to a second be the same as a distance on the *x*-axis corresponding to 2.9979 hundred thousand kilometers.

If we do this then the world line of a photon makes an angle of 45 degrees with the vertical, while anything moving with a speed *v* less than *c* has a world line making an angle θ of less than 45 degrees with the vertical (Fig. 17.5). It is useful to know the relation between the angle θ and the speed *v*. The tangent of θ (Fig. 17.6) is just the fraction of a light-second that the particle travels in 1 second, which, in turn, is given by the ratio of the particle's speed to the speed of light. Thus

$$\tan \theta = \frac{v}{c} .$$

We know that an observer B, moving to A's right with speed *v* along the *x*-axis, will have a coordinate system that differs from A's, since the two reach different conclusions about

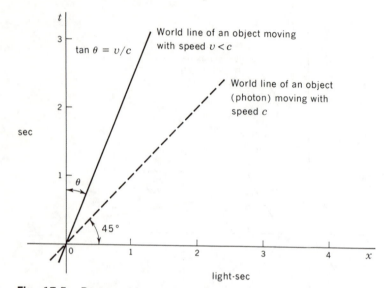

Fig. 17.5 Rectangular coordinates with a scale chosen so that the world line of a photon makes a 45-degree angle with the axes.

lengths, velocities, time intervals, and the the like. We wish to draw in A's diagram the set of space-time points making up B's x'- and t'-axes.* It is helpful, in doing this, to keep in mind

*This is the geometrical analog of calculating the Lorentz transformation equations, which express analytically the co-ordinates B gives an event in terms of the coordinates A assigns it.

a few general notions about coordinate systems and how they are used.

The coordinate system we have given to A is a familiar rectangular coordinate system, which locates points in space and time by superimposing on them a rectangular grid (Fig. 17.7). Two perpendicular lines* (drawn more heavily in Fig.

*Which two is arbitrary, corresponding to the freedom to choose which time we wish to call $t = 0$ and which place we wish to call $x = 0$.

17.7) are chosen to be the x- and t-axes. Horizontal lines are

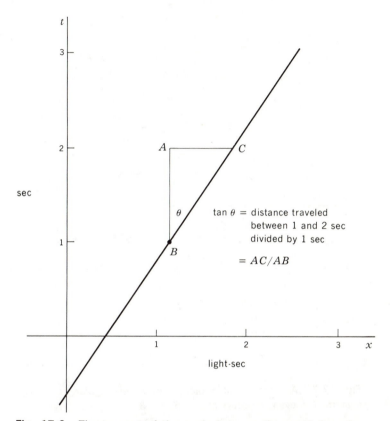

Fig. 17.6 The tangent of the angle between the world line of a uniformly moving particle and the vertical *t*-axis is just v/c, where v is the velocity of the particle in the rectangular coordinate system.

assigned numbers according to how far above the *x*-axis they lie, which are indicated in the scale along the *t*-axis. Similarly, vertical lines are assigned numbers according to how far to the right of the *t*-axis they lie, which are indicated in the scale along the *x*-axis. The *x*-coordinate of a point is just the number of the vertical line passing through it, and its *t*-coordinate is the number of the horizontal line passing through it. The point *P* in Fig. 17.7, for example, is at $x = 5$, $t = 3$.

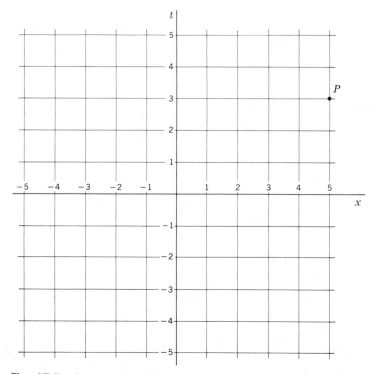

Fig. 17.7 A more detailed picture of a rectangular coordinate system. The event P occurs at $x = 5$, $t = 3$.

We belabor these facts to emphasize that nothing is essentially different if we do not use a rectangular grid. It is important to understand this because, as we shall see, if we give A a rectangular grid, B is forced to use an oblique coordinate system. This is a grid made up of two sets of parallel lines which, unlike A's, do not cross each other at right angles (Fig. 17.8). The lines of the grid are numbered just as they are in the rectangular system, the numbers being given in the scales along the t'- and x'-axes. The x'-coordinate of a point is the number associated with the line parallel to the t'-axis passing through the point, and the t'-coordinate is the number associated with the line parallel to the x'-axis passing through

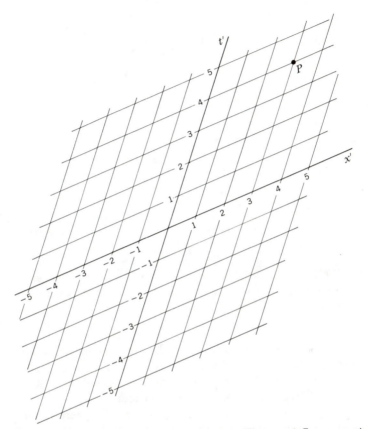

Fig. 17.8   An oblique coordinate system. The event P occurs at $x' = 3$, $t' = 4$.

the point. The point P in Fig. 17.8, for example, is at $x' = 3$, $t' = 4$.

If only the coordinate axes and their scales are drawn and not the full grid, we can find the coordinates of a point P as follows: Draw through the point two lines, one parallel to the x- (or x'-) axis and one parallel to the t- (or t'-) axis. The number appearing on the x- (or x'-) axis scale at the point where it is intersected by the line parallel to the t- (or t'-) axis is the x- (or x'-) coordinate of P, and the number appear-

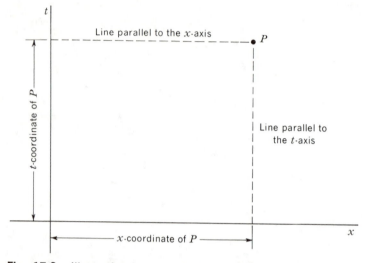

Fig. 17.9   Illustrating how one finds the x- and t-coordinates of a point P by parallel projection, when a grid like that in Fig. 17.7 is not given.

ing on the t- (or t'-) axis scale at the point where it is intersected by the line parallel to the x- (or x'-) axis is the t- or (t'-) coordinate of P. This sounds more complicated than it is. Figures 17.9 and 17.10 should make it clearer.

To specify an oblique coordinate system, we must specify the x'-axis, the t'-axis, and the scales on each. Since the x'-axis is the set of all points with t'-coordinate zero and the t'-axis is the set of all points with x'-coordinate zero, the things to be specified are the lines of constant t', the lines of constant x', and the scales.

A line of constant x' is the trajectory of a particle B considers to be stationary. Since B moves to A's right with velocity v, so will a particle stationary with respect to B. Therefore a line of constant x' in A's diagram is a straight line making an angle with the vertical whose tangent is v/c. Since B's t'-axis is the set of all space-time points having $x' = 0$, it is that particular line of constant x' for which the constant value of x' is zero. If we agree to set the origin of B's system so that he

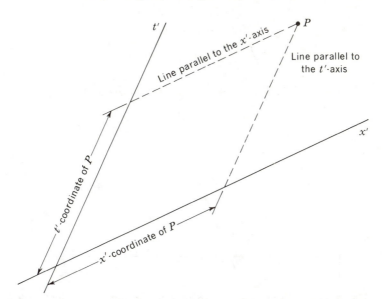

Fig. 17.10   Illustrating how one finds the x'- and t'-coordinates of a point P by parallel projection, when a grid like that in Fig. 17.8 is not given.

assigns the coordinates $x' = 0$, $t' = 0$ to the point A calls $x = 0$, $t = 0$,* the set of all points with $x' = 0$ includes in particular

> * This can always be done just by shifting without twisting the x'- and t'-axes. This again is due to the freedom we have in choosing which time and place to call $t' = 0$ and $x' = 0$.

the point $x = 0$, $t = 0$. Hence B's t'-axis is the straight line in A's diagram that passes through the origin and makes an angle θ with the t-axis, where $\tan \theta = v/c$. This is shown in Fig. 17.11.*

> * Note that when v is much less than c, θ is very near to zero and the t'-axis is almost indistinguishable from the t-axis. This is the price we have paid for compressing the x-axis and making it possible to distinguish speeds close to c from infinite speeds. It is now very hard to distinguish speeds much less than c from zero speed.

So far all we have done makes sense non-relativistically as well as relativistically. Relativistic effects first appear when we

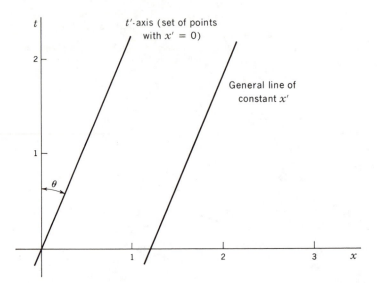

Fig. 17.11 World lines of two objects moving to the right with speed v with respect to the rectangular system ($\tan \theta = v/c$). For an observer moving with the objects (B) the lines are lines of constant x', and the one passing through the origin, the t'-axis.

look for the lines of constant t'. Non-relativistically the answer is that they are the same as lines of constant t, and therefore parallel to the x-axis. This cannot be so relativistically, since according to Rule 4, if A says that two spatially separated events happen at the same time, B will disagree. Thus lines of constant t' must differ from lines of constant t.*

> * For if A says two events are simultaneous, they lie on a line of constant t, but if B is to find they are not simultaneous, they cannot lie on a line of constant t'.

We can deduce what the lines of constant t' must be by exploiting the principle of the constancy of the velocity of light. Any observer must conclude that two stationary events occur simultaneously (i.e., lie on a line of constant t') if they are both illuminated by a flash of light that originated midway between them. Suppose, then, that B is on a train (which he of course considers to be at rest) moving to the right with speed v past A. At the center of the train a flashbulb goes off, photons from the flash traveling to the right and left with speed c. According to B the photons must reach the two ends of the train at the same time.

These events are shown in Fig. 17.12 *in A's diagram*. Suppose that at (A-time) $t = 0$, the train extends from D_0 to E_0, the flashbulb being at F_0 which is halfway from D_0 to E_0. The two ends of the train and the bulb all move to the right with speed v, and therefore all have parallel world lines making angles θ ($\tan \theta = v/c$) with the vertical. These world lines are just the slanting lines passing through D_0, F_0, and E_0.

At the space-time point F the bulb goes off, and the dashed lines FD and FE are the trajectories of the photons traveling

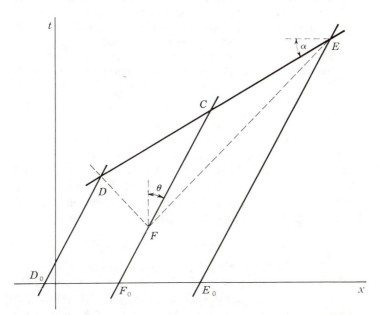

Fig. 17.12   If D_0, E_0, and F_0 are the ends and center of a moving train at A-time $t = 0$, then a flash of light at the center (event F) will reach the two ends (events D and E) simultaneously in a frame moving with the train. Hence DE is a line of constant t' for an observer with velocity v ($\tan \theta = v/c$). It follows from purely geometric reasoning that $\alpha = \theta$.

from the flash to the left and right. The dashed lines make angles of 45 degrees with the vertical because the speed of light is also c in A's frame.*

> * Note that this is where the constancy of the velocity of light comes in. We have now used in both frames the fact that the speed of light is c.

The photons reach the left end of the train at the space-time point D, and they arrive at the right end at the space-time point E. According to B, events D and E are simultaneous. Thus the line DE is a line of constant t'.*

> * Actually, of course, all we know is that the points D and E happen at the same B-time. However, the only relationship this analysis establishes between the two points is the angle the line joining them makes with the horizontal. Their absolute location and separation is arbitrary, being determined by the length of the train and when the bulb goes off. Hence any two points joined by a line making the same angle with the horizontal as the line joining D and E must be simultaneous in frame B. Since all points on the line through D and E are in this relationship, the line is a line of constant time, as asserted.

The determination of the angle DE makes with the x- or t-axis is now a problem in plane geometry. Since $D_0 D$, $F_0 F$, and $E_0 E$ are parallel lines and $D_0 F_0 = F_0 E_0$, it follows that $DC = CE$. This means that C bisects the hypotenuse of the right triangle DFE and is therefore equidistant from the vertices.* In particular, then, $CF = CE$ and hence angle $CFE =$

> * An angle inscribed in a semicircle is a right angle.

angle CEF. But the angle CF makes with the vertical (θ) plus angle $CFE = 45$ degrees, while the angle CE makes with the horizontal (α) plus angle $CEF = 45$ degrees. Therefore CE makes the same angle with the horizontal as CF makes with the vertical. But the angle CF makes with the vertical is just the angle θ, whose tangent is v/c. Therefore *lines of constant t' make the same angle $\theta = \tan^{-1}(v/c)$ with the horizontal as lines of constant x' make with the vertical.*

The x'-axis is that line of constant t' for which $t' = 0$. Since we have agreed to set the origin of B's coordinate system to coincide with that of A's, the set of all points with $t' = 0$ will include the origin. Thus the x'-axis is the line through the origin making an angle θ with the horizontal. We can therefore add some information to Fig. 17.11, as shown in Fig. 17.13.

All that now remain to be determined are the scales to be put on the x'- and t'-axes. We can reach a partial conclusion immediately: The length of t'-axis corresponding to a time (B-time) of 1 second must be the same as the length of x'-axis corresponding to a distance (B-distance) of 1 light-second (just as is the case for A-times, A-distance, and segments of the t- and x-axes). For consider the trajectory of a beam of light passing $x = 0$ at $t = 0$ (Fig. 17.14). A subsequent point on the trajectory P has coordinates x_0' and t_0' given by the intersection of the line of constant x', PB, with the x'-axis, and the line of constant t', PC, with the t'-axis. Since the speed of light is 1 light-second per second past B and since B finds that the light has traveled a distance x_0' in a time t_0', the segment of

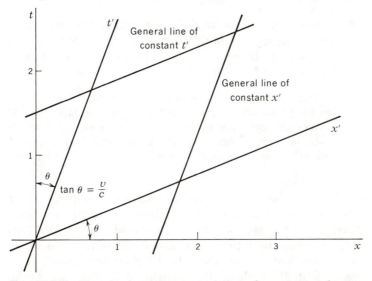

Fig. 17.13   The x'-axis and a general line of constant t' for an observer moving to the right with speed v are added to Fig. 17.11. The x'-axis makes the same angle θ with the horizontal as the t'-axis does with the vertical.

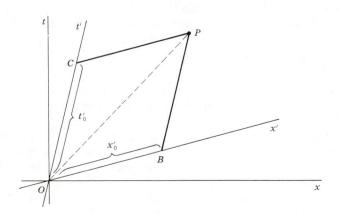

Fig. 17.14 A geometrical demonstration that a segment of x'-axis representing a distance of 1 light-second has the same length as a segment of $t' =$ axis representing a time of 1 second.

x'-axis from the origin to x_0', OB, must represent as many light-seconds as the segment of the t'-axis from the origin to t_0', OC, represents seconds. But all four sides of the parallelogram $OBPC$ are equal,* and therefore in particular the length OC

> * This follows from the fact that the diagonal OP bisects the angle COB, which in turn follows from the fact that OP makes 45-degree angles with the x- and t-axes, and from the fact, just established, that the angle between the x'- and x-axes is the same as the angle between the t'- and t-axes.

equals the length OB.

Therefore both A's and B's axes have the property that a piece of the spatial axis corresponding to a length of 1 light-second has the same length as a piece of the temporal axis corresponding to a time of 1 second. Thus the amount by which the scale on the x'-axis is stretched (or shrunk) compared with the scale on the x-axis must be the same as the corresponding amount for the t- and t'-axes. Since both pairs of axes have the same stretching factor, it is enough to deduce this factor for just one pair. We do this by invoking the principle of relativity.

Consider a stick 1 meter long, stationary in A's frame. If it extends from C to D along the x-axis at time $t = 0$, the space-time trajectories of its two ends are the vertical lines parallel to the t-axis passing through these points (Fig. 17.15).

An observer B moving to the right with velocity v will find (Fig. 17.16) that the stick extends along the x'-axis from E to F at time $t' = 0$.* Hence if a meter stick moving with velocity

> * It is important to understand clearly why this is so, for it contains the essential flavor of the Minkowski diagram approach. We must think of a meter stick, not as a single object, but as an immense collection of points (e.g., the 38.71282-centimeter mark, the 53.15193-centimeter mark, etc.), each of which has its own trajectory in space and time. Each trajectory, in turn, can be thought of as a continuum of points, a point being the particular piece of meter stick associated with the trajectory at a particular moment in its history. All the trajectories together fill the shaded region

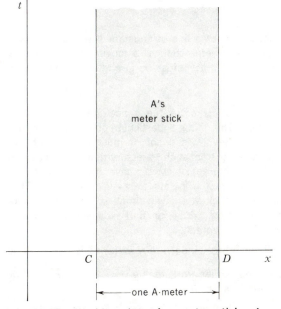

Fig. 17.15 World region of a meter stick, stationary in the rectangular system, that stretches from C to D at time zero.

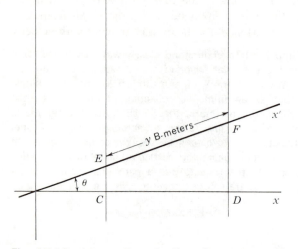

Fig. 17.16   Intersection of a line of constant t' (the x'-axis) with the set of points of A's meter stick. If a moving meter stick becomes y meters long, then EF must be a length of x'-axis representing y meters.

in Fig. 17.15 and make up the space-time entity that is the meter stick. The thing a particular observer calls "the meter stick at time t_0" is a set of meter-stick points assembled by choosing from each trajectory that point occurring at time t_0. Since different observers have different notions of simultaneity, the sets of points they pick will differ. Thus for A "the meter stick at a given time" is a horizontal slice of the shaded region, while for B it is a slice making an angle θ with the horizontal.

v parallel to its length has a length of y meters,* B's scale along

* We know from earlier work that $y = \sqrt{1 - v^2/c^2}$, but we want to deduce this again.

the x'-axis will be such that the distance EF corresponds to y meters.

Now the principle of relativity requires that if B has a rule that the true length of a moving stick is y times what an observer moving with the stick says, A must have the same rule. Hence, in particular, A must find that a stick moving with B and extending from E to F along the x'-axis has a length of $y \times y = y^2$ meters, since B says the length of such a stick is y meters. The trajectories of the ends of this second stick consist of lines through E and F making an angle θ ($\tan \theta = v/c$) with the vertical (since this stick moves with B), and hence at time (A-time) $t = 0$, A will find the stick extending along the x-axis from G to H (Fig. 17.17).

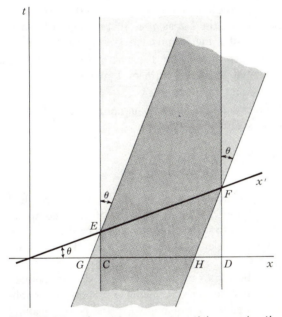

Fig. 17.17   A's stationary meter stick occupies the shaded region betwen the two vertical lines. B's stationary stick of length y meters occupies the shaded region between the two slanting lines, and therefore stretches along the x-axis from G to H at time $t = 0$. Since both observers have a rule that the length of a moving stick changes by a factor y, GH must represent an A-length of y^2 meters. Since CD represents an A-length of 1 meter, one can conclude geometrically that $y = \sqrt{1 - v^2/c^2}$, where $\tan \theta = v/c$.

Thus the length GH of x-axis is to be an A-length of y^2 meters. Since CD is an A-length of 1 meter, $GH/CD = y^2$. But we can also calculate this ratio as follows:

$GC = EC \tan \theta$;

$HD = FD \tan \theta$;

hence

$(FD - EC) \tan \theta = HD - GC = CD - GH$;

but

$FD - EC = CD \tan \theta$;

therefore from the last two lines, $\tan^2 \theta = 1 - GH/CD$.

Thus $y^2 = GH/CD = 1 - \tan^2 \theta = 1 - v^2/c^2$.

We have therefore proved from the diagrammatic analysis that $y = \sqrt{1 - v^2/c^2}$, reestablishing the Fitzgerald contraction. To establish the scale on the x'-axis, note that a length of x'-axis equal to a B-length of 1 meter must be $1/y$ times the length EF. But $EF = CD/\cos \theta = CD \sqrt{1 + v^2/c^2}$. Since CD is a length of x-axis equal to an A-length of 1 meter, *a length of x'-axis equal to a B-length of 1 meter is $\sqrt{(1 + v^2/c^2)/(1 - v^2/c^2)}$* $= 1/\sqrt{\cos 2\theta}$ *times a length of x-axis equal to an A-length of 1 meter.* The same scaling factor

$$\sqrt{\frac{1 + v^2/c^2}{1 - v^2/c^2}}$$

gives the factor by which the scale on the t'-axis is stretched compared with that on the t-axis, and therefore a complete description of the axes to be used by an observer B moving to the right past A with velocity v along the x-axis is as follows (see also Fig. 17.18):

Let the angle θ be $\tan^{-1}(v/c)$. Then the x'-axis is inclined at an angle θ to the x-axis (rising in the direction of B's motion), the t'-axis is inclined at an angle θ to the t-axis (leaning toward the direction of B's motion), and the scale on each axis is stretched by a factor $1/\sqrt{\cos 2\theta}$ from the scale of the x- and t-axes.

(We note, in passing, the following geometrical interpretation of the scaling factor: If we draw in A's coordinate system the hyperbola $x^2 - t^2 = 1$ (see Fig. 17.19), which intersects the x'-axis in a point P, then OP is just the segment of x'-axis whose B-length is 1 meter. For $OP^2 = AP^2 + OA^2$, and, since

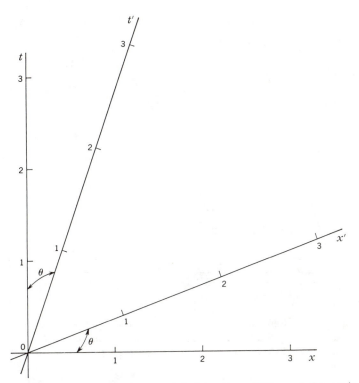

Fig. 17.18   Fig. 17.13, completed by the addition of the scale to the x'- and t'-axes.

P is on the hyperbola, $1 = OA^2 - AP^2$. Dividing the first of these relations by the second and taking the square root, we find $OP = \sqrt{(1 + \tan^2 \theta) / (1 - \tan^2 \theta)} = \sqrt{1/\cos 2\theta}$, since $AP/OA = \tan \theta = v/c$. In the same way, the hyperbola $t^2 - x^2 = 1$ intersects the t'-axis in a point which is 1 B-second along the t'-axis from the origin.)

As a first exercise in the use of these rules, we rederive the Lorentz transformation equations. Suppose an event happens at P (Fig. 17.20) to which A assigns the coordinates x, t ($OX = x, OT = t$). B gives the same event coordinates x' and t', where x' and t' are proportional to the lengths OX' and OT', the proportionality constant being $\sqrt{\cos 2\theta}$ (since B's scales are stretched by a factor $1 / \sqrt{\cos 2\theta}$). Hence*

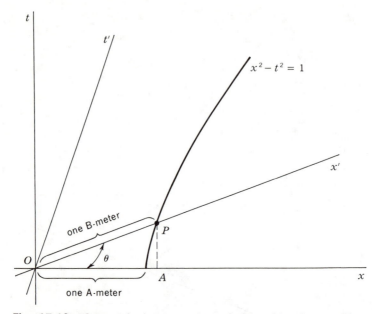

Fig. 17.19 Geometric interpretation of the scale change: The hyperbola $x^2 - t^2 = 1$ intersects the x'-axis of any observer at a point whose distance from the origin represents 1 meter for that observer.

$$x = OX, \qquad t = OT, \tag{17.1}$$

$$x' = OX' \sqrt{\cos 2\theta}, \qquad t' = OT' \sqrt{\cos 2\theta}. \tag{17.2}$$

* We are assuming here that the scale factor that converts the lengths OX and OT in the diagram into A's measurements x and t, is unity. This is permissible because the only relevant quantity in relating x', t' to x, t is the ratio of this scale factor to the scale factor B uses in going from the lengths OX' and OT' in the diagram to x' and t'. The ratio between these two scaling factors is just the factor giving the stretching of the primed axes compared with the unprimed ones, $1/\sqrt{\cos 2\theta}$. In other words if we replaced (17.1) by $x = aOX$, $t = aOT$, where a was a scaling number (for example 1 light-second per centimeter of page), the same a would appear on the right sides of each of Eqs.

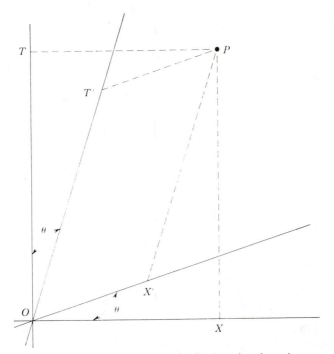

Fig. 17.20 The coordinates in both primed and un-primed systems of an event P.

(17.2) as an extra factor. In the Lorentz transformation equations, which involve only the ratios x/x' and t/t', these extra factors cancel out.

Note that

$$OT' = X'P. \qquad (17.3)$$

$$OT = XP. \qquad (17.4)$$

and hence all the lengths we wish to relate appear in the piece of Fig. 17.20 reproduced in Fig. 17.21. Applying the law of sines to the triangle PBX' we find

$$\frac{X'P}{BP} = \frac{\sin(90° + \theta)}{\sin(90° - 2\theta)} = \frac{\cos\theta}{\cos 2\theta} \qquad (17.5)$$

Furthermore

$$BP = XP - OX\tan\theta = OT - OX\tan\theta. \qquad (17.6)$$

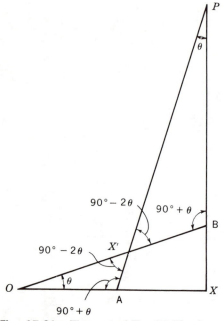

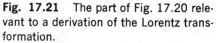

Fig. 17.21 The part of Fig. 17.20 relevant to a derivation of the Lorentz transformation.

Equations (17.3) through (17.6) require that

$$OT' = \frac{\cos \theta}{\cos 2\theta} (OT - OX \tan \theta). \tag{17.7}$$

Using (17.1) and (17.2) to replace the distances on the graph by the actual measurements of A and B, we convert (17.7) into

$$t' = \frac{\cos \theta}{\sqrt{\cos 2\theta}} (t - \tan \theta x). \tag{17.8}$$

Since $\tan \theta = v/c$ and $\cos 2\theta = \cos^2 \theta - \sin^2 \theta$, (17.8) can also be written

$$t' = \frac{t - vx/c}{\sqrt{1 - v^2/c^2}}. \tag{17.9}$$

We derive a similar relation between x', t, and x by considering the triangle OAX'. The law of sines gives

$$\frac{OX'}{AO} = \frac{\sin (90° + \theta)}{\sin (90° - 2\theta)} = \frac{\cos \theta}{\cos 2\theta}. \tag{17.10}$$

From Fig. 17.21 we also have

$$AO = OX - XP \tan \theta = OX - OT \tan \theta. \tag{17.11}$$

Combining these gives

$$OX' = \frac{\cos \theta}{\cos 2\theta} (OX - OT \tan \theta), \tag{17.12}$$

which gives, when we convert from distances on the graph to measured quantities,

$$x' = \frac{\cos \theta}{\sqrt{\cos 2\theta}} (x - \tan \theta t) \tag{17.13}$$

or, since $\tan \theta = v/c$,

$$x' = \frac{x - vt/c}{\sqrt{1 - v^2/c^2}}. \tag{17.14}$$

Equations (17.9) and (17.14) differ from the Lorentz transformation Eqs. (13.10) and (13.13) because the former set applies only to our diagrams, which use units in which the speed of light is unity. If we continue to measure time in seconds but wish to measure lengths in kilometers rather than light-seconds, we must use the fact that x kilometers are only x/c light-seconds, where c is 2.9979 hundred thousand kilometers per second. Hence if x and x' are given to us in kilometers, we must divide each by c to convert them into light-seconds before we can use (17.9) and (17.13). If we do this (17.9) becomes

$$t' = \frac{t - v(x/c)/c}{\sqrt{1 - v^2/c^2}},$$

or

$$t' = \frac{t - vx/c^2}{\sqrt{1 - v^2/c^2}}, \tag{17.15}$$

and (17.14) becomes

$$\frac{x'}{c} = \frac{x/c - vt/c}{\sqrt{1 - v^2/c^2}},$$

or

$$x' = \frac{x - vt}{\sqrt{1 - v^2/c^2}}. \tag{17.16}$$

Equations (17.15) and (17.16) are now exactly the same as the Lorentz transformation equations.*

> * They of course reduce back to (17.9) and (17.14) if x and x' are measured in light-seconds, since the speed of light is 1 light-second per second in such units, and therefore extra or missing factors of c make no difference, c being equal to 1.

Notice that Eqs. (17.14) and (17.9) are more symmetric than the Lorentz transformation Eqs. (17.15) and (17.16). The only difference between (17.14) and (17.9) is that the spatial and temporal coordinates are interchanged. The full beauty of the Lorentz transformation only emerges when we use units in which the speed of light is 1. Because nature has furnished us with an absolute velocity, it is not necessary to establish conventional units of *both* distance and time, for given the unit of distance, the "natural" unit of time is the time it takes light to travel that distance, and given a unit of time, the natural unit of distance is the distance light can travel in that time. Our reward for using natural units is that equations simplify and hidden symmetries emerge. The Minkowski diagrams, drawn with natural units, fully display the symmetry between space and time, the spatial and temporal axes of moving observers tilting and stretching in precisely the same way.

We can now illustrate various relativistic conundrums in Minkowski diagrams, thereby exposing their basic simplicity.

    1. *Graphical explanation of Rule 4.* Figure 17.22 shows two clocks, synchronized and separated by a distance l in their proper frame, which is taken to be the unprimed or A system (lines of constant time horizontal). For an observer B moving to the right with speed v along the x-axis, lines of constant time (such as the x'-axis) make angles θ with the horizontal, where $\tan \theta = v/c$. The moving observer checks the synchronization of the clocks by comparing their readings at the same time (B-time). Graphically, this means that he compares their readings at two points along their world lines that are intersected

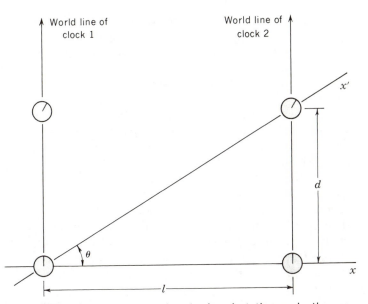

Fig. 17.22 Two clocks, synchronized and stationary in the rectangular frame, are shown at two different moments. A moving observer has tilted lines of constant time and therefore says the two clock configurations lying on the x'-axis are simultaneous.

by the same line of constant t' (the particular line of constant t' chosen in Fig. 17.22 is the x'-axis). Since B's lines of constant t' are different from the lines of constant t used by A in the proper frame of the clocks, he will certainly not agree that they are synchronized. Indeed, it is evident from the figure that according to B the left-hand clock (according to him, the clock in front, since he moves to the right) is behind the right-hand one.

The amount of the disparity is easily derived from the diagram. Since B compares the clocks at the same t', he compares the picture of clock 1 on the x-axis not with the picture of clock 2 on the x-axis (as A would do in a synchronization check), but with the picture of clock 2 on the x'-axis. From the point of view of the proper frame of the clocks, an amount of time given by the length d has elapsed between B's observations of the two clocks, where $d = l \tan \theta = lv/c$. Since clock

2 is stationary in A's frame it measures A-time, and has therefore advanced by just this amount between B's observations.
B therefore concludes that clock 2 is ahead of clock 1 (or
clock 1 is behind clock 2) by lv/c, which becomes lv/c^2 if we
measure l in kilometers rather than light-seconds, since 1 kilometer is $1/c$ light-second.

Thus A can attribute B's conclusion that the clocks are unsynchronized to the fact that B compared them using tilted
lines of constant time or, more concretely, that B compared
them using his own clocks which, from A's point of view, were
actually the unsynchronized ones. To see that things are quite
symmetric in this respect, note Fig. 17.23, which gives the
diagrammatic representation of A's analysis of B's clocks. Since
these move, their world lines are tilted at an angle θ to the
vertical, and since they are synchronized in their proper frame
(frame B) they have the same reading along lines of constant t'.
When A compares the clocks he slices their world lines with a

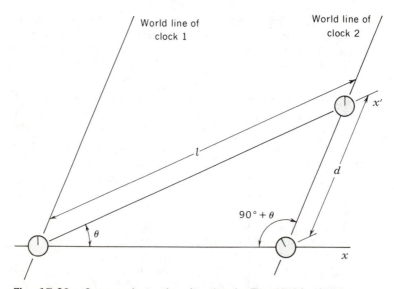

Fig. 17.23 Conversely to the situation in Fig. 17.22, if B says
two of his clocks are synchronized (i.e., read the same where they
cross the x'-axis) then they will certainly not read the same when
they intersect the x-axis.

horizontal line (the x-axis in the diagram) and therefore finds the clock in front (2) behind the clock in the rear (1). To find the disparity d, note that from the law of sines, $d/l = \dfrac{\sin \theta}{\sin (90° + \theta)}$. But the right side of this equation is just $\sin \theta / \cos \theta = v/c$. Therefore we again have $d = lv/c$, or, in conventional units, $d = lv/c^2$.

Note that in this very basic analysis we did not even need to know the scaling factor between the two coordinate systems. This is because the final result involved the ratio of a length (the proper distance between the clocks) to a time (the disparity between the clocks at two moments the moving observer says are simultaneous, which, since the clocks measure their own proper time, is the proper time between these moments) where both the length and the time are measured in the *same* system (the proper frame of the clocks).

2. *Graphical explanation of the reciprocity of the Fitzgerald contraction.* The Minkowski diagram in Fig. 17.24 shows the trajectories in space-time of two meter sticks in relative motion along a line parallel to their lengths. Stick A is stationary in the unprimed (x, t) system. Only the trajectories of its ends are shown in the diagram, as the two vertical lines. The locations of the marks 10, 20, . . ., 90 centimeters are shown at $t = 0$, that is, at the points where their trajectories intersect the x-axis, but their world lines are not drawn, to keep the figure from becoming too complicated. They would be, however, simply the set of vertical lines passing through these points.

Stick B is a meter stick drawn moving to the right at a speed $v = 3c/5$. Consequently the world lines of its two ends are the two lines making an angle $\tan^{-1} (3/5)$ with the vertical. The scale, in steps of 10 centimeters, is drawn on stick B at time $t' = 0$ by indicating where these points intersect the x'-axis. The world lines of these points, if drawn in the figure, would be lines parallel to the world lines of the two ends of the stick.

The reciprocity of the Fitzgerald contraction is evident from the picture. At time $t = 0$ B's stick extends (along the x-axis) from the zero to the 80-centimeter marks of stick A, and therefore an observer A concludes that it has shrunk by a factor of

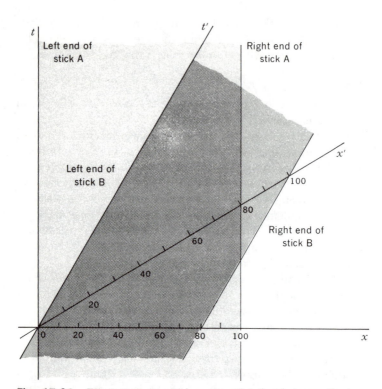

Fig. 17.24   The trajectory of A's meter stick lies between the two vertical lines, and the trajectory of B's meter stick, between the two slanted lines. B says that at time zero both sticks extend along the x'-axis and A says that at time zero both sticks extend along the x-axis. Thus each thinks the other's stick is only 80 centimeters long.

4/5, which is just $\sqrt{1 - (3/5)^2}$. In the same way, at time $t' = 0$ A's stick extends (along the x'-axis) from the 0- to the 80-centimeter marks of stick B. Thus observers A and B each justly concludes that the others' stick is shorter. The diagrammatic explanation can be put somewhat clumsily into words, as follows:

Both observers agree that A's stick is located on the set of space-time points between the two vertical lines, while B's is located on the set of space-time points between the two slanted lines. In other words, if no axes were drawn, but simply the two sticks, both A and B would agree that A's stick filled the lightly shaded region (Fig. 17.25), while B's filled the more darkly shaded region. Their disagreement is on how the lines of constant time are to be put into the picture.

But it is precisely the lines of constant time that determine what A or B *means* by "the stick." For the notion of the stick includes implicitly the assumption that all the points of matter making up the stick exist at the same moment. Thus the stick at a given moment is a slice of Fig. 17.25 with a line of constant time. Slicing the figure with horizontal lines of constant time gives the sticks at a given moment according to A's coordinates, and slicing it with a line tilted upward at an angle

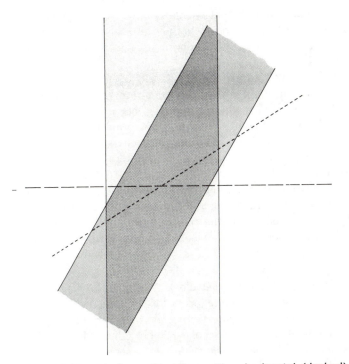

Fig. 17.25   A slices space-time with a horizontal (dashed) line, while B prefers to do it with a slanted (dotted) line. As a consequence each asserts that the other's stick is longer.

$\tan^{-1}(3/5)$ gives the sticks at a given moment according to B's coordinates. A slice of the first kind (dashed line) clearly leads to the conclusion that B's stick is shorter than A's, but a slice of the second kind (dotted line) equally clearly implies that A's stick is shorter than B's. There is no contradiction simply because a different definition of "A's stick" and "B's stick" appears in the two statements. The relativistic thinker must always remember that "A's stick" really means "the set of all points taken from the set of all trajectories of pieces of A's stick at the same moment," and "at the same moment" means different things to A and to B.

3. *Graphical explanation of the reciprocal slowing down of moving clocks.* Figure 17.26 shows two clocks in relative motion, with velocity $3c/5$. Clock A is stationary in the unprimed rectangular system at $x = 0$; therefore its trajectory lies along the t-axis. Clock B moves to the right; therefore its world line defines the t'-axis of a primed system in which it is stationary.

At the moment the two clocks pass each other, both read 0, as can be seen in the figure. Suppose an observer moving with clock A is asked what clock B reads 5 seconds after the encounter. He will, of course, interpret this to mean "what is the reading of clock B *at the same time* that clock A reads 5 seconds." He will therefore answer the question by slicing the diagram with a line of constant time (horizontal for A observers) passing through the world line of clock A at the moment clock A reads 5. This line (see Fig. 17.26) slices the world line of clock B at the moment clock B reads 4, and A therefore concludes that clock B reads 4 seconds when his reads 5, and hence that it is running slowly by a factor $\sqrt{1 - (v/c)^2} = 4/5$.

B, of course, if asked what clock A reads when *his* reads 5 seconds, follows the same procedure, slicing the diagram with a line of constant t' intersecting the trajectory of *his* clock at the moment it reads 5 seconds. Such a line (Fig. 17.26) slices the world line of clock A at 4 seconds, so B concludes that A's clock runs slowly by the same factor $4/5$.

The possibility that each can conclude that the other's clock runs slower is due to the fact that a comparison of two spatially

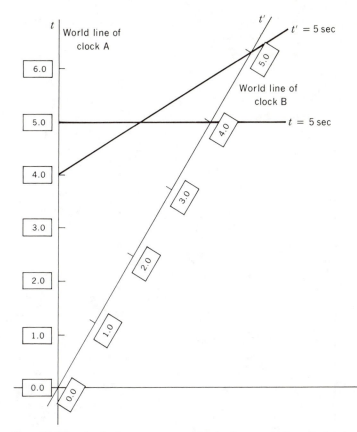

Fig. 17.26 A clock stationary with A shown at 1-second intervals, and one moving with B shown at 1-second intervals. Each observer can conclude that the other's clock is running slowly, since the comparisons are made along different lines of constant time.

separated clocks cannot be made until one has a notion of simultaneity. Since A and B use different lines of constant time, it is to be expected that they reach different conclusions.

   4. *The clock paradox.* All the analysis of Chap. 16 can be summarized in a few diagrams. Figure 17.27 shows the world lines of an outgoing shuttle (B_1) leaving the Earth (rectangular frame) at a speed of $4c/5$ and of an incoming shuttle (B_2)

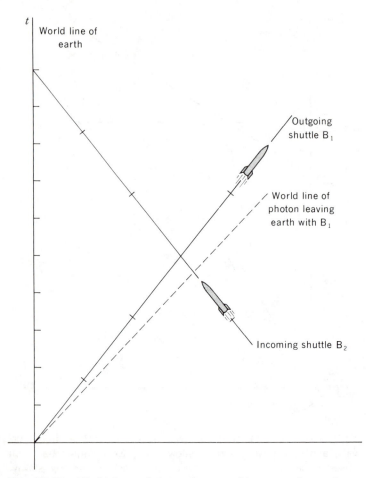

Fig. 17.27 World lines of the rockets used by a traveler such as B of Chap. 16. Marks on the *t*-axis are one earth-year apart. Marks on the rocket trajectories are one rocket-year apart.

returning to Earth at the same speed. The world line of the Earth is taken as the *t*-axis, and the marks on it represent intervals of 1 year. Thus 10 years elapse on Earth between the departure of B_1 and the arrival of B_2. (The particular numbers used here differ from those in Chap. 16, which are not as convenient for the diagrammatic analysis.)

The scale markings on the trajectories of the two ships represent times of 1 year, according to clocks on the ships. Note that they are farther apart than the scale markings on the t-axis by a factor of about 2.13. They were arrived at by noting that, for example, the world line of ship B_1 is the time axis of a frame in which the clocks of B_1 are stationary. In such a frame these clocks measure correct time, and therefore the length of axis during which the clocks advance by 1 year is related to the length of t-axis during which the Earth clocks advance 1 year by the scaling factor $1 / \sqrt{\cos 2\theta}$. When $\tan \theta = 4/5$, this works out to be $\sqrt{41}/3$, which is about 2.13.

Note that the outgoing shuttle meets the incoming one just 3 years (according to its own clocks) after it leaves Earth and that another 3 years pass on the incoming shuttle before it reaches Earth. Thus a man who rides out on one shuttle and in on the other experiences 6 years of time, while his colleagues on Earth have lived for 10 years when he returns. This is consistent with the fact that, from Earth's point of view, the moving man's clocks have run slowly by a factor $\sqrt{1 - (4/5)^2} = 3/5$ throughout his journey.

Figure 17.28 illustrates the fact that from the traveler's point of view the clocks on Earth run slowly throughout the journey. Just before he leaves frame B_1 he finds the reading of clocks on Earth by slicing the diagram with a line of constant time (B_1-time) passing through the world line of shuttle B_1 at the moment the clocks on the ship read 3 years. Note that this line of constant time intersects the t-axis at 1.8 years (which is, as expected, three-fifths of 3 years). Immediately after changing ships B can again check what the time is on Earth by slicing the diagram with a line of constant time passing through the point of transfer, but this time, being on shuttle B_2, he uses a line of constant B_2-time. This slopes upward to the left* and

---

* Note that diagrams considered up to now have all had observers moving to the right from the point of view of the rectangular frame. The rules for coordinate systems moving to the left are exactly the same, except that θ is replaced by $-\theta$, which is natural, since v is replaced by $-v$. Hence the angle between the axes of an observer moving to the left expands to an obtuse angle, rather than contracting to an

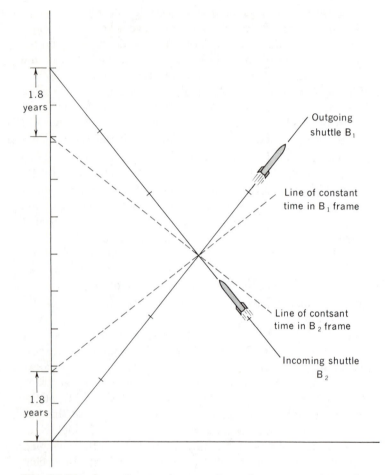

1.8 years

Outgoing shuttle B$_1$

Line of constant time in B$_1$ frame

Line of contsant time in B$_2$ frame

Incoming shuttle B$_2$

1.8 years

Fig. 17.28 According to observers B$_1$ at the moment of transfer, 1.8 years have passed on earth since B's departure, but according to observers B$_2$ at the moment of transfer, all but 1.8 of the full 10 years have past. The disparity is due to the two observers slicing the diagram with different lines of constant time.

acute angle, as it does for an observer moving to the right. In particular the axes of an observer moving to the left are such that lines of constant x are tilted at an angle θ from the vertical to the *left*, and lines of constant time are tilted at an angle θ *down* from the horizontal (as one moves to the right along the line). This is illustrated in Fig. 17.29.

meets the t-axis at a point 1.8 years before the end of the journey. The missing $10 - 1.8 - 1.8 = 6.4$ years are, in a sense, swept out by the changing line of constant time as B changes ships.

The analysis in terms of what B *sees* is pictured in Fig. 17.30. The dashed lines are trajectories of light signals given off once a year on Earth. When $v/c = 4/5$ the rate at which these signals are received on the ship is every $\sqrt{(1 + v/c)/(1 - v/c)}$ $= 3$ years on the way out and every $\sqrt{(1 - v/c)/(1 + v/c)}$ $= \frac{1}{3}$ year on the way in (as discussed in the Appendix to Chap. 5). Thus flashes are seen on the outgoing ship every 3 years and on the incoming ship three times a year. This analysis

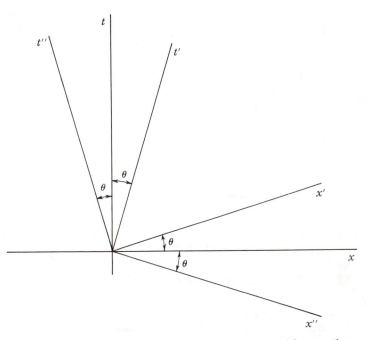

Fig. 17.29 An observer moving to the right with speed v ($\tan \theta = v/c$) uses the x'-t' system of axes. An observer moving to the left with speed v uses the x''-t'' system.

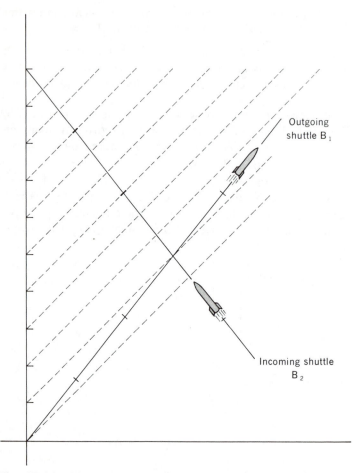

Fig. 17.30 The trajectories of photons (dashed lines) from the annual flashes on Earth set off during B's journey. One occurs at his send-off, the next reaches him at the moment of transfer, and eight more reach him on the return journey. The last (not shown) occurs upon his return.

is confirmed by the diagram. The first flash from Earth arrives at B_1 3 years after departure, just as B prepares to transfer. On the remaining 3 years' journey home he sees nine more flashes and therefore observes all the 10 years that passed on Earth.

On the other hand (Fig. 17.31) B's annual flashes are seen on the Earth with the same frequencies: every 3 years for flashes emitted from the outgoing ship and every third of a year for flashes emitted from the incoming ship. Observers on

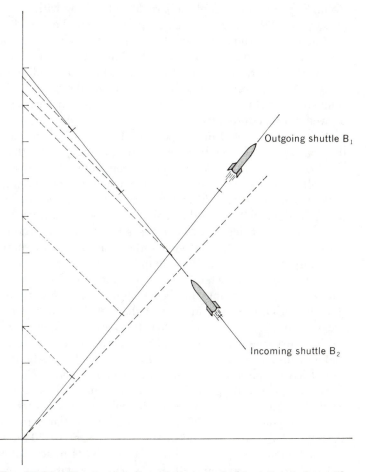

Outgoing shuttle B$_1$

Incoming shuttle B$_2$

Fig. 17.31 The trajectories of photons (dashed lines) from B's annual flashes. The three emitted up to, and including, the moment of transfer arrive on Earth 3 years apart. The next three, emitted up to, and including, the moment of arrival, arrive every 4 months.

Earth therefore take 9 years to see the 3 year's worth of out-going flashes and then see the remaining 3 years of incoming flashes in a single year.

5. *Another "paradox."* Minkowski diagrams are very helpful in resolving superficially inconsistent views of a single event. Suppose, for example, a barn 40 meters long with a door at each end lies along the *x*-axis. Let both doors be closed. A man now comes running down the *x*-axis from the left, holding up a horizontal pole of proper length 50 meters, parallel to the *x*-axis. The man runs at four-fifths the speed of light, and so the Fitzgerald contraction reduces the length of the pole in the frame of the barn to a mere 30 meters. The following sequence of events is therefore possible:

When the right end of the pole reaches the left barn door, the door opens to allow the man and his pole to enter without slowing down. As soon as the pole is completely in the barn, the left door closes again. The pole continues on its way, traveling the additional 10 meters before its right end reaches the right door, with both doors shut. As soon as its right end reaches the right door, that door opens to let the pole out and does not close until the left end of the pole has left the barn. The important aspect of these humdrum events is that since the pole is only 30 meters long in the frame of the 40-meter long barn, there is a time during which the pole is completely shut up within the barn, in spite of the fact that when at rest the pole is 10 meters longer than the barn. In other words we have exploited the Fitzgerald contraction to enclose completely within a container a moving object which, if stationary, would have been too big to fit inside the container.

All this is perfectly possible, but at first glance it looks somewhat alarming in the rest frame of the pole. For in the runner's frame the pole is a full 50 meters long, while the barn has shrunk to 24 meters in length. How, then, can the 50-meter pole be momentarily enclosed in a barn only 24 meters long?

The answer is, of course, that it is not. The statement that the pole is enclosed in the barn depends on the fact that the left door closes before the right door opens. But this fact is based on a judgment of the order in time of two spatially separated events, which sometimes can, and in this case does, change, depending on which observer makes the judgment. This is easily seen in the Minkowski diagrams of Figs. 17.32 and 17.33. The two vertical lines are the trajectories of the two barn doors, drawn solid when the doors are closed and dashed when open. The two parallel slanted lines are the trajectories of the two ends of the stick. The situation at various important moments in the frame of the barn is shown in Fig. 17.32. Since

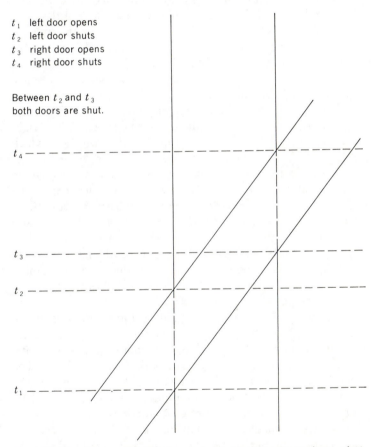

t_1 left door opens
t_2 left door shuts
t_3 right door opens
t_4 right door shuts

Between t_2 and t_3
both doors are shut.

t_4

t_3

t_2

t_1

Fig. 17.32 The vertical lines each represent a barn door, shut when solid, and open when dashed. The slanted lines are the world lines of two ends of a pole. The horizontal dashed lines are several lines of constant time in the frame of the barn.

we are describing things in the frame of the barn, we find the configuration at a particular time by slicing the diagram with a horizontal line of constant time. Thus at time t_1 the pole is just about to enter the barn. Prior to t_1 the pole is to the left of the barn and both doors are shut. After t_1 the left door opens and the stick enters the barn. At time t_2 the stick is completely in the barn and the left door closes. Between t_2 and t_3 the stick crosses the interior of the barn with both doors closed. At time t_3 the right end of the stick reaches the right door of the barn, which opens. Between t_3 and t_4 the stick passes through the open right door of the barn, which closes at t_4, after which the stick moves away to the right of the barn.

To describe these events in the frame of the stick, we must now slice the diagram with lines of constant time that make an angle $\tan^{-1}(4/5)$ with the horizontal. This is done in Fig. 17.33. At time t_1 the situation is still as described earlier: The right end of the stick has just arrived at the left barn door, which opens. The next significant event is that the right end of the stick reaches the right end of the barn at time t_2, whereupon the door at that end opens. Note that at t_2 the left end of the stick has not yet entered the left end of the barn. Between times t_2 and t_3 the stick extends beyond the doors of the barn on both the left and right sides, and both barn doors are open. This situation persists until t_3, at which moment the left end of the stick finally enters the barn, and the left door is closed. Somewhat later at t_4, the left end of the stick leaves the barn and both doors are closed again.

Thus in the frame of the stick, at no time is more than one barn door closed while any part of the stick is in the barn, even though in the frame of the barn there is a period during which the stick is in the barn and both doors are closed. The moral is that even a harmless statement like "the stick is in the barn" contains an implicit assumption of simultaneity which, if forgotten, can lead to apparent contradictions.

It is worth emphasizing again that, as in the case of the two meter sticks in Fig. 17.25, there is a basic reality that both observers agree on. This is pictured in Fig. 17.34. What the barn and stick observers disagree on is how the lines of constant time are to be laid across the diagram. When the con-

t_1 left door opens
t_2 right door opens
t_3 left door shuts
t_4 right door shuts

Between t_2 and t_3
both doors are open.

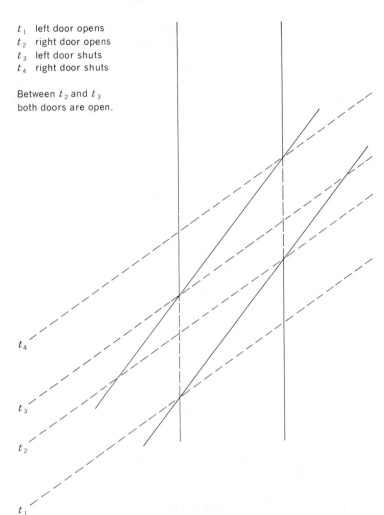

Fig. 17.33   The same as Fig. 17.32 except that now the lines of constant time are taken in the frame of the pole, and are therefore tilted.

troversy is put that baldly, it is clear that there is no paradox at all. The great virtue of Minkowski diagrams is that they enable us, by drawing a picture of our colloquial verbal description of a situation, to get at this underlying reality. Once

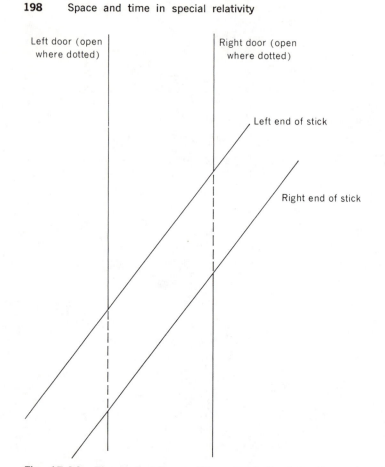

Fig. 17.34 The underlying reality, unprejudiced by lines of constant time.

we have the picture in any frame in which it is simple to draw, we can immediately see how things must be described in any other frame simply by tilting the space and time axes according to the Minkowski rules.

Before concluding this chapter, I should mention the extension of the Minkowski-diagram technique to events not lying on a straight line. When the events lie in a plane, the diagrams become three dimensional, for it is necessary to add a y-axis, perpendicular to both the x- and t-axes. An observer moving with velocity v along the x-axis will describe events using an x'- and t'-axis which are precisely as we described them above, lying in the x-t plane, and, in addition, he will have a y'-axis perpendicular to the x'-t' plane (and hence also perpendicular to the x-t plane) that is parallel to the y-axis and has the same scale. Thus nothing unusual happens to the axis perpendicular to the direction of motion. This follows from Rules 1 and 5.

If we wish to describe motion taking place in all three spatial dimensions, the utility of the Minkowski diagram all but vanishes, for it is then necessary to introduce a z-axis, perpendicular to the x-t plane and to the y-axis. Such a state of affairs (two axes perpendicular to the same plane and to each other) cannot be produced in even a 3-dimensional diagram and would require a diagram in a 4-dimensional space* for its com-

* This is none other than that spooky fourth dimension of early post-relativistic pop science fiction.

plete representation. At this point, unless one's sense of 4-dimensional geometry is unusually keen, one is better off either thinking of an analogous problem involving spatial motion in only one or two dimensions or reverting to the analytic approach of the Lorentz transformation equations.

18

ENERGY AND MOMENTUM

The aim of this book has been to describe and explain the behavior of moving clocks and meter sticks, and from this point of view a chapter on energy and momentum is not relevant. It is included for two reasons, only one of which is compelling. The bad reason is that $E = mc^2$ has become so familiar a formula that an elementary book on relativity must say something about it. More important, however, is the fact that the direct consequences of the relativistic laws of energy and momentum conservation provide the most abundant and convincing experimental justification of the entire theory. The design of any high-energy particle accelerator requires an engineer who is completely familiar with special relativity; a machine built by a non-relativistic engineer will not work.

Unfortunately this chapter cannot possess the degree of rigor maintained in the earlier ones. The conservation of energy and momentum is a principle stating that certain quantities dependent on the velocities of a group of particles at a given moment retain the same values at different moments in the history of those particles, even though as a result of the forces they exert on one another, the particles may be moving in complicated ways. We should properly begin by listing all the forces particles can exert on one another, such as electromagnetic forces, nuclear forces, and gravitational forces. We should then have to develop theories of the behavior of particles interacting through these forces before we could extract a law of energy and momentum conservation.

Such a program would require several additional books, and I shall avoid it by giving arguments which are plausible, but not nearly as convincing. We shall first of all discuss only collisions (which are often of chief interest in practical problems). A collision is an encounter between particles which begins and ends with several particles all very far apart from one another, well beyond the reach of their forces of interaction, and hence* moving uniformly in straight lines. During

> * Newton's first law of motion.

the encounter they may bump into one another, spin around one another, or otherwise interact, as long as at the very beginning and very end there are a number of particles moving uniformly.* We shall take as an experimental fact the con-

> * For the moment we shall assume that the same particles are present at the beginning and at the end of the encounter, i.e., that there is no fission or fusion of the particles. Eventually this restriction will be dropped.

servation of ordinary non-relativistic energy and momentum when all the particles are moving at speeds much less than c. We shall then try to *guess* what a relativistic definition of energy and momentum should be by requiring that:

1. When all the particles have speeds much less than c, the relativistic energy and momentum are indistinguishable from their non-relativistic values.

2. If any inertial observer calculates the energy and momentum of a group of particles and finds they are conserved in a collision, so will any other inertial observer.

The first requirement is simply that the relativistic conservation laws reproduce known facts, and the second stipulates that these conservation laws not be accidental consequences of a particular observer having a special velocity with respect to the particles. Remarkably enough, these requirements are enough to specify the form energy and momentum must take in a relativistic theory.*

> * When I referred to the lack of rigor in this chapter, I had in mind two shortcomings. We shall not establish here that energy and momentum *must* be conserved, but only the form they must have *if* they are conserved. Furthermore, we shall find a form consistent with requirements (1) and (2) but shall not show that it is the only one (although I challenge you to try to find another).

With these reservations in mind, we can proceed with the argument. We first review the non-relativistic definitions of energy and momentum. Suppose a particular observer finds that a uniformly moving object has coordinates x_0, y_0, z_0 at time t_0, and then at time t_1, coordinates x_1, y_1, z_1. We define the x, y, and z components of the velocity of the object by

$$u_x = \frac{x_1 - x_0}{t_1 - t_0},$$

$$u_y = \frac{y_1 - y_0}{t_1 - t_0},$$

$$u_z = \frac{z_1 - z_0}{t_1 - t_0}. \tag{18.1}$$

The magnitude of the velocity u (also known as the speed) of the object is the total distance d it has gone divided by the time it took; therefore

$$u = \frac{d}{t_1 - t_0} = \frac{\sqrt{(x_1 - x_0)^2 + (y_1 - y_0)^2 + (z_1 - z_0)^2}}{t_1 - t_0}$$

$$= \sqrt{u_x^2 + u_y^2 + u_z^2}. \tag{18.2}$$

The non-relativistic energy* of a particle of mass m is de-

> * Strictly speaking only the kinetic energy, but if we are only considering particles out of the range of all forces, then we can choose the potential energy to be zero.

fined as

$$E = \tfrac{1}{2}mu^2, \tag{18.3}$$

and the three components p_x, p_y, and p_z of its non-relativistic momentum, by

$$p_x = mu_x, \qquad p_y = mu_y, \qquad p_z = mu_z. \tag{18.4}$$

The total energy of a group of particles of masses m^1, m^2, etc., and speeds u^1, u^2, etc., is just the sum of their individual energies:

$$E = \Sigma E^i = \Sigma \tfrac{1}{2} m^i \, u^{i2}.$$
$$(\Sigma E^i \equiv E^1 + E^2 + \dots), \qquad \text{etc.} \tag{18.5}$$

The total x, y, and z components of the momentum are similarly defined:

$$P_x = \Sigma m^i \, u_x{}^i, \qquad P_y = \Sigma m^i \, u_y{}^i, \qquad P_z = \Sigma m^i \, u_z{}^i. \tag{18.6}$$

The law of conservation of energy and momentum is the statement that the total energy and three components of the total momentum have the same value before the collision as they do after, in spite of the changes of the energies and momenta of the individual particles.

Note that if energy and momentum are conserved for one non-relativistic observer, they are conserved for all non-relativistic observers.* This is easily verified by using the non-

> * This is the requirement we shall later impose on the relativistic definition of energy and momentum.

relativistic rule for the addition of velocities. Suppose the quantities in (18.5) and (18.6) are the energy and momenta of the particles as measured by A, and suppose B moves with velocity v along the x-axis of A. Then if A says the velocity of the ith particle has components $u_x{}^i$, $u_y{}^i$, $u_z{}^i$, B will find that its components are

$$u_x{}^{i\prime} = u_x{}^i - v, \qquad u_y{}^{i\prime} = u_y{}^i, \qquad u_z{}^{i\prime} = u_z{}^i. \tag{18.7}$$

Hence B will find that the energy and momentum of the ith particle are

$$\begin{aligned}
E^{i\prime} &= \tfrac{1}{2} m^i [(u_x{}^i - v)^2 + u_y{}^{i2} + u_z{}^{i2}] \\
&= E^i - m^i \, u_x{}^i \, v + \tfrac{1}{2} m^i \, v^2, \\
p_x{}^{i\prime} &= m^i (u_x{}^i - v) = p_x{}^i - m^i v, \\
p_y{}^{i\prime} &= p_y{}^i; \qquad p_z{}^{i\prime} = p_z{}^i.
\end{aligned} \tag{18.8}$$

Adding these individual energies and momenta to get the total energy and momentum, B finds

$$
\begin{aligned}
E' &= \Sigma E^{i\prime} = \Sigma E^i - v\,\Sigma m^i\,u_x{}^i + \tfrac{1}{2}v^2\,\Sigma m^i \\
&= E - P_x v + \tfrac{1}{2}Mv^2, \\
P_x' &= \Sigma p_x{}^{i\prime} = \Sigma p_x{}^i - v\,\Sigma m^i = P_x - Mv, \\
P_y' &= P_y, \qquad P_z' = P_z,
\end{aligned}
\tag{18.9}
$$

where M is the total mass of the group of particles, $M = \Sigma m^i$.

Evidently since A and B agree on the numerical values of the total y and z components of the momentum, if A says each has the same value before and after the collision, so will B. Furthermore, since $P_x' = P_x - Mv$ and M and v are the same before and after the collision,* if P_x has the same value before

---

* M is the total mass, which is certainly a constant non-relativistically, and v is the relative velocity of A and B, which has nothing to do with what the particles are doing.

---

and after the collision, so will P_x'. Finally, since $E' = E - P_x v + Mv^2/2$, if E and P_x both have the same values before and after the collision, so will E'.

This completes the non-relativistic proof that if A finds energy and momentum to be conserved, so will B.

We can verify immediately that this is not the case for an observer moving with relativistic velocities with respect to A and hence establish that (18.3) and (18.4) are not good definitions of energy and momentum in a relativistic theory. This is simply because the non-relativistic velocity addition law (18.7) no longer holds. Consider a particle found by A to be at x_0, y_0, z_0 at time t_0. Using the Lorentz transformation (Eqs. (13.10)–(13.13)), we find that according to B the particle will be at

$$
x_0' = \frac{x_0 - vt_0}{\sqrt{1 - v^2/c^2}}, \qquad y_0' = y_0, \qquad z_0' = z_0'
\tag{18.10}
$$

at time

$$
t_0' = \frac{t_0 - vx_0/c^2}{\sqrt{1 - v^2/c^2}}.
\tag{18.11}
$$

Similarly, the moment in the history of the particle that A says occurs at x_1, y_1, z_1 at time t_1, B says occurs at

$$x_1' = \frac{x_1 - vt_1}{\sqrt{1 - v^2/c^2}}, \qquad y_1' = y_1, \qquad z_1' = z_1 \qquad (18.12)$$

at time

$$t_1' = \frac{t_1 - vx_1/c^2}{\sqrt{1 - v^2/c^2}}. \qquad (18.13)$$

B calculates the components of the velocity of the particle using the ordinary definition

$$u_x' = \frac{x_1' - x_0'}{t_1' - t_0'}, \qquad u_y' = \frac{y_1' - y_0'}{t_1' - t_0'}, \qquad u_z' = \frac{z_1' - z_0'}{t_1' - t_0'}.$$
$$(18.14)$$

When combined with (18.10) to (18.13) this gives

$$u_x' = \frac{x_1 - x_0 - v(t_1 - t_0)}{t_1 - t_0 - (v/c^2)(x_1 - x_0)},$$

$$u_y' = \sqrt{1 - \frac{v^2}{c^2}} \frac{y_1 - y_0}{t_1 - t_0 - (v/c^2)(x_1 - x_0)},$$

$$u_z' = \sqrt{1 - \frac{v^2}{c^2}} \frac{z_1 - z_0}{t_1 - t_0 - (v/c^2)(x_1 - x_0)}. \qquad (18.15)$$

If we divide numerator and denominator of the right side of each of these equations by $t_1 - t_0$ and use (18.1), we find the relativistic generalization of (18.7) (we omit the superscript i that identifies the particular particle):

$$u_x' = \frac{u_x - v}{1 - u_x v/c^2}, \qquad u_y' = \sqrt{1 - \frac{v^2}{c^2}} \frac{u_y}{1 - u_x v/c^2},$$

$$u_z' = \sqrt{1 - \frac{v^2}{c^2}} \frac{u_z}{1 - u_x v/c^2}. \qquad (18.16)$$

The first of these is just the addition law for parallel velocities discussed in Chap. 14. The remaining two equations are the generalization of this result to components of the velocity that are not parallel to the relative velocity of A and B. Note also that (18.16) reduces to (18.7) in the case $v \ll c$, as it must.

You will discover if you try to use (18.16) instead of (18.7) that the proof that the non-relativistic energy and momenta are conserved for B if they are conserved for A can no longer be constructed. The trouble is that with B's velocities related to A's by (18.16), B's expression for the *total* momentum and energy of a group of particles is a complicated function of A's expressions for the *separate* momenta and energies of each of them. If we assume A's expression for the *total* energy and momentum has the same value before and after the collision, since B's expression for the total quantities depends on more than just these, A's conservation laws no longer imply conservation laws for B.

This suggests that to discover the correct form for the relativistic energy and momentum, we should look for quantities that transform more simply under a change of observer than the components of the velocity. For example the coordinates themselves transform simply as (18.10) and (18.11) or (18.12) and (18.13) show. These may not at first appear to be simpler than (18.16), but they are, for one important reason: In (18.10) and (18.11) the unprimed quantities appear only in the numerators, but in (18.16) they appear in numerator *and* denominator. This means that although the primed and unprimed coordinates are related by a set of linear equations with constant coefficients, the primed and unprimed velocities are not so related relativistically.*

* Although they were, non-relativistically, and in a particularly simple way (Eq. (18.7)).

The reason the relativistic transformation equations for the velocity are so much more complicated is that in calculating the velocity, the difference in the two position coordinates is divided by the difference in the two time coordinates, which itself changes as observers are changed. If instead, we divided by a quantity that all observers gave the same numerical value to (a so-called invariant), the quantity we arrived at would transform in no more complicated a way than the coordinates themselves, under a change of inertial frames. Furthermore, if we could choose this invariant so that it reduced to the time difference in the non-relativistic limit, the quantities we constructed would reduce to the ordinary velocity at non-relativistic speeds. Hence, multiplying it by m, we would have a candidate for the relativistic momentum which satisfied condition 1—that it reduce to the non-relativistic momentum in the non-relativistic region.

The invariant that replaces the time interval $t_1 - t_0$ between the two events x_1, y_1, z_1, t_1 and x_0, y_0, z_0, t_0 is simply the time interval T between the two events as measured by an observer with respect to whom they happen in the same place or, what amounts to the same thing, as measured by a clock traveling with the particle. This is called the "proper time" interval between the two events.* Since non-relativistically all observers

> * All observers agree on the proper time interval between two events, since the definition of the proper time interval specifies which frame the measurement is to be made in (namely the frame in which the particle is at rest). Alternatively, it can be related to the reading of a specified clock between two specified events.

agree on temporal measurements, this quantity reduces to the ordinary time interval in the non-relativistic limit. However in general an observer A who finds the particle to be moving with a speed u will find that the time (A-time) between two points on the particle's trajectory is related to the proper time by

$$t_1 - t_0 = \frac{T}{\sqrt{1 - u^2/c^2}}, \qquad (18.17)$$

since A finds that a clock moving with the particle runs slowly.

If we define a generalization of the ordinary velocity (known as the 4-velocity, for reasons that will presently become clear) by

$$w_x = \frac{x_1 - x_0}{T}, \qquad w_y = \frac{y_1 - y_0}{T}, \qquad w_z = \frac{z_1 - z_0}{T}, \quad (18.18)$$

then using (18.17) to eliminate T, we find

$$w_x = \frac{x_1 - x_0}{t_1 - t_0} \frac{1}{\sqrt{1 - u^2/c^2}},$$

$$w_y = \frac{y_1 - y_0}{t_1 - t_0} \frac{1}{\sqrt{1 - u^2/c^2}},$$

$$w_z = \frac{z_1 - z_0}{t_1 - t_0} \frac{1}{\sqrt{1 - u^2/c^2}}, \qquad (18.19)$$

or

$$w_x = \frac{u_x}{\sqrt{1 - u^2/c^2}}, \qquad w_y = \frac{u_y}{\sqrt{1 - u^2/c^2}},$$

$$w_z = \frac{u_z}{\sqrt{1 - u^2/c^2}}. \tag{18.20}$$

It is clear from (18.20) that when u is much less than c, the components of w are, as expected, almost identical to the components of u.

Now from the definition (18.18) it is evident* that B's com-

* Since T is a quantity all observers have the *same number* for.

components of w are related to A's in exactly the same way that B's coordinates are related to A's, i.e., by a Lorentz transformation like (18.10) and (18.11) (or (18.12) and (18.13)) except that x_0, y_0, and z_0 are replaced by w_x, w_y, and w_z. But that is not quite all, since the time coordinate t_0 gets mixed up with the spatial ones under a Lorentz transformation, and so we must also have a w_t to put into the transformation equations in place of t. From a glance at (18.18) it is apparent that w_t must be defined by

$$w_t = \frac{t_1 - t_0}{T}, \tag{18.21}$$

which (18.17) tells us is just

$$w_t = \frac{1}{\sqrt{1 - u^2/c^2}}. \tag{18.22}$$

If we multiply these four components of w (whence the name "4-velocity") by the mass m of the particle (which we assume is also a quantity that all observers agree on),* we pro-

* One sometimes runs across statements that, on the contrary, the mass of a particle is not an invariant quantity. This is a semantic problem that we will attend to shortly.

duce a set of four quantities which still transform under a change of observer, just like the coordinates and time. We call these quantities p_x, p_y, p_z, and p_t, defining the relativistic momentum to be given by the first three. We shall see what to do with p_t in a moment. Thus

$$p_x = \frac{mu_x}{\sqrt{1 - u^2/c^2}},$$

$$p_y = \frac{mu_y}{\sqrt{1 - u^2/c^2}},$$

$$p_z = \frac{mu_z}{\sqrt{1 - u^2/c^2}},$$

$$p_t = \frac{m}{\sqrt{1 - u^2/c^2}}. \tag{18.23}$$

Because of the way in which they were constructed, if B, moving with velocity v along A's x-axis, calculates the same quantities using the velocities *he* finds the particles to have, his values will be related to A's by the Lorentz transformation Eqs. (18.10) and (18.11) with x_0, y_0, z_0, t_0 replaced by p_x, p_y, p_z, p_t:*

$$p_x' = \frac{p_x - vp_t}{\sqrt{1 - v^2/c^2}},$$

$$p_y' = p_y,$$

$$p_z' = p_z,$$

$$p_t' = \frac{p_t - vp_x/c^2}{\sqrt{1 - v^2/c^2}}. \tag{18.24}$$

* If you found the argument leading to (18.24) a bit too delicate to be convincing, you should verify as an exercise in algebra that if A defines p_x, etc., by (18.23), while B defines p_x', etc., by (18.23) but with every component of u replaced by the corresponding component of u', then (18.24) follows directly (but somewhat complicatedly) from the Eqs. (18.16) that relate the components of u and u'.

Equations (18.23) define a set of four quantities that transform simply (i.e., by linear equations) under a change of frames, three of which reduce to the three components of the non-relativistic momentum, when u is very much less than c. We shall show presently that with this definition of the relativistic momentum, the conservation laws hold in all frames, but there are first two unfinished matters to be dealt with: we still need a quantity that reduces to the non-relativistic energy in the appropriate limit, and it would also be interesting to know what, if anything, the significance of p_t might be. Here one of those happy economies of physics occurs.

Suppose we try to decide what p_t is by looking at its non-relativistic limit, i.e., by precisely the same method we used to identify the spatial components of the momentum. At first glance this appears to be uninteresting, since when u is much less than c, p_t is very close to m, the mass of the particle, a fixed number, independent of both the velocity of the particle and the observer. This, however, is only because we have not looked closely enough; $p_t - m$ is not strictly zero unless u is precisely zero. In general it can be written this way:

$$p_t - m = m\left(\frac{1}{\sqrt{1 - u^2/c^2}} - 1\right) = m\left(\frac{1 - \sqrt{1 - u^2/c^2}}{\sqrt{1 - u^2/c^2}}\right)$$

$$= \frac{mu^2/c^2}{\sqrt{1 - u^2/c^2}\,(1 + \sqrt{1 - u^2/c^2})}$$

$$= \tfrac{1}{2}mu^2\left(\frac{1}{c^2}\frac{1}{\sqrt{1 - u^2/c^2}}\frac{2}{1 + \sqrt{1 - u^2/c^2}}\right). \qquad (18.25)$$

Now when u/c is much smaller than 1, $\sqrt{1 - u^2/c^2}$ is very close to 1, the quantity in parentheses in the last line of (18.25) is very close to $1/c^2$, and $p_t - m$, very close to $\tfrac{1}{2}mu^2/c^2$, that is, to $1/c^2$ times the non-relativistic kinetic energy! If we multiply both sides of (18.25) by c^2, the non-relativistic limit is

$$p_t c^2 = mc^2 + \tfrac{1}{2}mu^2, \qquad u \text{ much less than } c, \qquad (18.26)$$

which says that $p_t c^2$ reduces in the non-relativistic limit to the kinetic energy plus a constant. This constant is, of course, enormously bigger than the kinetic energy, which is why we did not notice the extra $\tfrac{1}{2}mu^2$ after a first superficial glance. But an important property of the non-relativistic energy is that one can add to it any arbitrary numerical constant without changing its significance.*

* This is because the only thing of physical significance is the *difference* in the energies of two configurations, from which the constant cancels out.

The simplest thing we could do therefore is to define the relativistic energy as $p_t c^2$, which takes care of both problems at once, telling us what to do with p_t and giving us a relativistic energy that reduces to the non-relativistic expression when u is much less than c. In addition, if the relativistic energy is so defined, it too will transform in a simple way under a change of observers.

In this case the simplest answer is also the correct one (a phenomenon that has occurred with extraordinary frequency throughout the history of physics):*

$$E = p_t c^2 = \frac{mc^2}{\sqrt{1 - u^2/c^2}}.$$ (18.27)

* One immediate and pleasing consequence of this choice is that it provides a dynamical explanation for why nothing can go faster than light, for as u approaches c, the energy of the particle grows without bound. Therefore no matter how large an amount of energy we are willing to supply to a particle, its final velocity will still be only (solving (18.27) for u in terms of E) $u = c\sqrt{1 - (mc^2/E)^2}$, which for large E/mc^2 is very close to, but still less than, the speed of light.

The formulas we have been led to for the energy and momentum of a particle moving with relativistic velocity are summarized below:

If the components of the particle's velocity are u_x, u_y, and u_z and the mass of the particle is m,

$$E = \frac{mc^2}{\sqrt{1 - u^2/c^2}},$$

$$p_x = \frac{mu_x}{\sqrt{1 - u^2/c^2}},$$

$$p_y = \frac{mu_y}{\sqrt{1 - u^2/c^2}},$$

$$p_z = \frac{mu_z}{\sqrt{1 - u^2/c^2}}.$$ (18.28)

If E, p_x, p_y, p_z are the energy and momentum as measured by A, the energy and momentum measured by B moving with velocity v along the x-axis of A will be E', p_x', p_y', p_z' where

$$E' = \frac{E - vp_x}{\sqrt{1 - v^2/c^2}},$$

$$p_x' = \frac{p_x - vE/c^2}{\sqrt{1 - v^2/c^2}},$$

$$p_y' = p_y, \qquad p_z' = p_z. \tag{18.29}$$

(These results follow immediately from (18.23) and (18.24) after (18.27) is used to express p_t in terms of E.)

We can conclude directly from (18.29), which describes how the energy and momentum of a single particle transform under a change of observers, that if one observer finds that the total energy and momentum of a group of particles is conserved, so will any other. All we must do to demonstrate this is to put a superscript ^{i} on all the E', p', E, and p occurring in (18.29) to signify the ith particle and then sum the result over all i:

$$\Sigma E^{i\prime} = \frac{1}{\sqrt{1 - v^2/c^2}}\left(\Sigma E^i - v\Sigma p^i\right),$$

$$\Sigma p_x{}^{i\prime} = \frac{1}{\sqrt{1 - v^2/c^2}}\left(\Sigma p_x{}^i - \frac{v}{c^2}\Sigma E^i\right),$$

$$\Sigma p_y{}^{i\prime} = \Sigma p_y{}^i, \qquad \Sigma p_z{}^{i\prime} = \Sigma p_z{}^i. \tag{18.30}$$

If we now define the total energy E by

$$E = \Sigma E^i$$

and the total momentum P by

$$P_x = \Sigma p_x{}^i,$$
$$P_y = \Sigma p_y{}^i,$$
$$P_z = \Sigma p_z{}^i,$$

(18.30) can be written as

$$E' = \frac{E - vP_x}{\sqrt{1 - v^2/c^2}}, \qquad P_x' = \frac{P_x - vE/c^2}{\sqrt{1 - v^2/c^2}}$$

$$P_y' = P_y, \qquad\qquad P_z' = P_z, \tag{18.31}$$

which is the same as (18.29), except that the quantities appearing in (18.31) are now the *total* energy and components of the *total* momentum.* But it is now evident, just as it was

> * It is this simplicity of the transformation properties of the *total* energy and *total* momentum that makes it so desirable to have linear transformation laws for the energy and momentum of single particles.

in the analogous non-relativistic argument, that if A says that the total energy and momentum of a group of particles are the same before and after a collision, B must also find they are the same, since (18.31) shows that B's values are uniquely determined by A's.

Thus the definitions (18.28) satisfy both requirements we set out to impose on candidates for relativistic energy and momentum, and we shall take them to be the correct ones. One point, however, deserves more attention. Consider the formula for the energy of the *i*th particle

$$E^i = \frac{m^i c^2}{\sqrt{1 - u^{i2}/c^2}}. \tag{18.32}$$

We noted that this reduced to the non-relativistic kinetic energy plus a constant, when u was much less than c. On this basis alone we could equally well conjecture that the energy of a particle of mass m^i should be

$$E^i = \frac{m^i c^2}{\sqrt{1 - u^{i2}/c^2}} + e^i, \tag{18.33}$$

where e^i is an additional constant, characteristic of the particle (like its mass) and independent of its velocity. Evidently (18.33) also reduces to

$$E^i = \tfrac{1}{2}m^i u^{i2} + \text{constant},$$

when u^i is much less than c. Furthermore our other guiding criterion, that energy be conserved for all observers, continues to be satisfied under the definition (18.33). For if the initial group of particles is the same as the final group, all the definition (18.33) does is to add the same extra term Σe^i to both the initial and final energies.

The only objection we can offer against (18.33) is that the transformation equations (18.31) become rather more complicated:

$$E' = \frac{E - vP_x}{\sqrt{1 - v^2/c^2}} - \left(\frac{1}{\sqrt{1 - v^2/c^2}} - 1 \right) \Sigma e^i,$$

$$P_x' = \frac{P_x - vE/c^2}{\sqrt{1 - v^2/c^2}} + \frac{v/c^2}{\sqrt{1 - v^2/c^2}} \Sigma e^i,$$

$$P_y' = P_y,$$

$$P_z' = P_z. \tag{18.34}$$

This extra complexity arises only in the relativistic case. If we change the non-relativistic definition (18.3) to

$$E^i = \tfrac{1}{2} m^i \, u^{i2} + e^i,$$

the non-relativistic transformation law (18.9) remains precisely the same. There is thus a slight incentive to set all the e^i equal to zero in the relativistic theory, which is completely lacking in the non-relativistic case:

Relativistic energy and momentum transform under a change of observers in the simplest way if the additive constant in the energy is chosen so that the energy of a particle of mass m when at rest is $E = mc^2$.* In contrast, the non-relativistic trans-

---

* This is just another way of saying $e^i = 0$, as is evident from setting u equal to zero in (18.33).

---

formation laws are the same no matter what value is assigned to the additive constant in the energy.

Relativity having provided us with a reason, however fragile, for making a particular choice of the constant in the energy, we shall follow the time-honored method and make the simplest choice, resorting to a more complicated one only if forced to by a contradiction between theory and experiment.

However, having taken this daring stand, we are immediately faced with the embarrassing fact that experiment cannot refute us: if the same group of particles is present before and after the collision, conservation of energy is unaffected by adding a constant to the definition of a particle's energy. Our choice has therefore been a vacuous one.

This is unfortunate. When there is a reason, however slight, for making a choice between otherwise equivalent alternatives, it is always worthwhile exploring the consequences of that choice, and disappointing when there are no consequences at all. In the present case, however, there is a way out, although it is rather a bold one. The extra constant in the initial energy is the same as the extra constant in the final energy only if the initial group of particles is the same as the final group of particles. If we therefore require that energy be conserved even in collisions in which the final group of particles is not the same as the initial group, i.e., in collisions during which particles merge together, split in half, etc., then if energy is conserved for some choice of the e^i, it will not necessarily be conserved for any other, and making the choice $e^i = 0$ acquires physical content.

We shall therefore make another generalization of the non-relativistic law of energy conservation, defining the relativistic energy by (18.27) *and* requiring that energy as so defined be conserved in *all* collisions, whether or not the initial group of particles is the same as the final group. This choice turns out to be abundantly confirmed by experiment (and required by more sophisticated theories).*

> * Let me emphasize that the argument preceding this generalization was in no way meant to be compelling, only mildly suggestive. My point was that relativity does provide us with a motivation, absent non-relativistically, for giving a particular value to the energy constant, and this choice acquires physical content only if energy conservation is extended to collisions in which the nature or number of particles is not conserved.

From a non-relativistic point of view this generalization is quite unexpected. Non-relativistically, total *momentum* is conserved whether or not particles merge or split apart in the course of a collision, but total kinetic energy is not. This is simply seen in the case of two soft clay balls of equal mass m and opposite velocities of magnitude u which collide head on, forming a single ball of mass $2m$ and zero velocity. Total momentum is conserved (being zero before and after the collision) but total kinetic energy is not (being mu^2 before and zero, after).*

> * Total energy of all kinds is, however, conserved. The missing kinetic energy is converted into thermal energy: the

bigger ball is somewhat warmer. This production of thermal energy also occurs for relativistic balls of clay, but the con-servation of energy as defined in (18.32) now implies, as will be seen below, that this thermal energy produces an extra contribution to the mass of the composite ball.

Note that this non-relativistic asymmetry between energy and momentum does not prevent momentum in such collisions from being conserved for all observers. For the non-relativistic transformation laws (18.9) give the primed momentum as a function *only* of the unprimed momentum, even though they give the primed energy as a function of both the unprimed momentum *and* the unprimed energy. Therefore conservation of momentum in one frame is enough to guarantee it in all frames non-relativistically, even when energy is not conserved.

On the other hand the relativistic transformation laws (18.31) give the primed momentum as a function of the unprimed momentum *and* the unprimed energy, and therefore relativisti-cally conservation of momentum in one frame is *not* enough to guarantee it in all frames unless energy is also conserved.

This fact lends compelling support to our extension of en-ergy conservation to collisions in which the number and nature of the particles changes (sometimes known as inelastic colli-sions). If we had not extended conservation of energy to apply to this case relativistically, then automatically conservation of momentum would also not have applied. Thus our choice was not as radical as it may have appeared: If we had not chosen to *extend* the law of energy conservation in going to the rela-tivistic case, we should have been forced to *restrict* the law of momentum conservation. Happily, in this case the correct choice is the one that broadens, not diminishes, the range of physical law.

Let us now consider the case of the two colliding clay balls, using the relativistic conservation law. If the final com-posite ball has mass M and is stationary, its energy is Mc^2, and the energy of each of the balls before it collides is $mc^2/\sqrt{1-(u/c)^2}$. Since energy is conserved

$$Mc^2 = \frac{2mc^2}{\sqrt{1-u^2/c^2}}.$$

If we write this as

$$M = 2m + 2m \left(\frac{1}{\sqrt{1 - u^2/c^2}} - 1 \right), \qquad (18.35)$$

we see that the mass of the composite ball exceeds the masses of the two balls we started with. If we define the energy of motion of an object as its energy minus its energy when at rest, i.e., its energy minus mc^2, the mass of the composite ball exceeds the total mass of the two original balls by precisely their energy of motion divided by c^2.

Note that in the case $u \ll c$, this takes the simple form*

$$M = 2m + \frac{1}{c^2} 2 \left(\tfrac{1}{2}mu^2 \right). \qquad (18.36)$$

* Since $mc^2 / \sqrt{1 - (u/c)^2} = mc^2 + \tfrac{1}{2}mu^2$ to a high degree of accuracy, when u is much less than c.

Thus the kinetic energy that was lost non-relativistically is restored relativistically as an enhancement in the mass of the final ball. This is true quite generally. Kinetic energy that is lost (i.e., converted to other forms of energy such as heat) in a non-relativistic theory is not lost in a relativistic theory, because the masses of the final particles are enhanced by just the missing kinetic energy divided by c^2. From this point of view the question that should worry us is not why energy is conserved even in inelastic collisions in the relativistic theory, but why it was thought not to be conserved in inelastic collisions non-relativistically. The answer to the latter question is quite simple:

When u is much less than c,

$$\Sigma E^i = \Sigma m^i c^2 + \tfrac{1}{2}\Sigma m^i u^{i2}, \qquad (18.37)$$

and this quantity is indeed conserved in all collisions, elastic or inelastic, as long as all the speeds of the particles are very small compared with c, so that the approximation (18.37) remains accurate. The non-relativistic theory errs, however, in its assumption that the sum of the initial masses equals the sum of the final masses even in inelastic collisions. This assumption, known as the law of conservation of mass, is a very natural mistake to make non-relativistically, because the deviations from conservation of mass involve only fractional changes in mass of amounts comparable to $(u/c)^2$ where u is a typical particle velocity (compare (18.36)). In the non-relativistic case u/c is much less than 1, and hence the changes in mass involved are minute, and almost impossible to detect. If, for example, the speeds of the initial clay balls were that of a fast jet airplane, the mass of the composite ball would not be the sum of the initial masses, $2m$, but something of about the size 2.0000000001 m. On the other hand the changes in classical kinetic energy (from mu^2 to zero) are quite easy to detect. Thus of the two terms on the right side of (18.37), changes in the first are almost impossible to detect when non-relativistic processes are involved, but changes in the second are very easy to measure. It was therefore not realized that changes were present in the first term that precisely canceled changes in the second, leading to conservation of relativistic energy, even in inelastic collisions.*

> * Indeed, it was not realized that the first term was present at all, but merely writing it down would not have dispelled the non-relativistic error as long as conservation of mass was believed in.

Some of the most striking confirmations of special relativity come from the observation of just such changes in the behavior of atomic nuclei. If a radioactive nucleus of mass M decays while at rest into two particles of masses M_1 and M_2, for total relativistic energy to be conserved, the two particles must have velocities u_1 and u_2 satisfying

$$M = \frac{M_1}{\sqrt{1 - u_1^2/c^2}} + \frac{M_2}{\sqrt{1 - u_2^2/c^2}},$$

in order that energy be conserved. Thus the loss of mass* must

> * If the sum of the masses of the decay products were greater than the initial mass, no choice of velocities could lead to conservation of energy.

appear in the form of kinetic energy of motion. This prediction has been abundantly confirmed: The total energy of motion of the decay products is just the difference between the initial and final total masses times c^2.

The nuclear case is particularly suited for verifying this law, since the nuclear masses are known with great precision and, more important, the velocities of the final particles are large enough that the diminution in mass can be appreciable. However, such a loss of mass occurs whenever internal energy (nuclear, electrical, chemical, etc.) is converted into energy of motion. Only in the nuclear case is the amount of energy converted so large that when divided by the enormous number c^2 it still gives an observable change in mass, but in principle $E = mc^2$ is as descriptive of a chemical explosive, a gasoline engine, or a flying bird as it is of a nuclear explosion.

The converse case, in which the mass of a composite particle is less than the sum of the masses of the particles composing it, is even more common. If you look through a table of nuclear weights, you will discover that every nucleus weighs less than the total weight of the neutrons and protons out of which it is constructed. This mass deficit times c^2 turns out to be precisely the amount of energy that has to be supplied to the nucleus in order to disassemble it into the appropriate number of separate neutrons and protons. The same thing is true of chemically bound systems: A molecule of water weighs less than two hydrogen atoms and an oxygen atom, but by an undetectably small amount. The disparity in the nuclear case is greater, because the nuclear forces are much stronger than the electrical ones (which are responsible for chemical binding), and hence more energy is required to pry apart a nucleus than a molecule.

For practical purposes it is often convenient to replace the four Eqs. (18.28) defining the energy and momentum of a particle of mass m by the algebraically equivalent set:*

$$E^2 - c^2(p_x^2 + p_y^2 + p_z^2) = (mc^2)^2,$$

$$\frac{cp_x}{E} = \frac{u_x}{c}, \qquad \frac{cp_y}{E} = \frac{u_y}{c}, \qquad \frac{cp_z}{E} = \frac{u_z}{c}. \qquad (18.38)$$

* Yet another way to do it, now considered old fashioned, is to define a "relativistic mass" by $\mu = m/\sqrt{1 - u^2/c^2}$. Equations (18.28) then become $E = \mu c^2$, $p_x = \mu u_x$, $p_y = \mu u_y$, $p_z = \mu u_z$. People who do things this way talk about a velocity dependent mass μ, which becomes larger as u gets closer

to c. Evidently this is just a matter of language, depending on whether you want the factor $\sqrt{1 - u^2/c^2}$ to go with the u, or with the m. Most people choose the factor to go with the u, reflecting the fact that it arises through $p_x = mw_x$, $p_y = mw_y$, $p_z = mw_z$, and $w_x = u_x / \sqrt{1 - u^2/c^2}$, $w_y = u_y / \sqrt{1 - u^2/c^2}$, $w_z = u_z / \sqrt{1 - u^2/c^2}$.

Another very important result is that $E^2 - c^2 (P_x^2 + P_y^2 + P_z^2)$ is a quantity on which all observers agree, where E is the total energy and the P's, the components of the total momentum of a group of particles. This follows directly from the transformation rules (18.31), which imply that

$$E'^2 - c^2 (P'_x{}^2 + P'_y{}^2 + P'_z{}^2) = E^2 - c^2 (P_x^2 + P_y^2 + P_z^2).$$

(18.39)

The first of Eqs. (18.38) is an example of this general rule in the case where the group is just a single particle of mass m. The fact that all observers agree on the value they give to $E^2 - c^2 (p_x^2 + p_y^2 + p_z^2)$ there becomes the fact that they all assign the same mass to the particle.

We conclude this chapter with a practical example of the very different kinds of conclusions a relativistic engineer must draw compared with his non-relativistic colleagues. Relativistic engineers like to manufacture new particles out of energy. They do this by the primitive method of hurling a very energetic particle at a stationary one and waiting to see what happens after the collision. The best that could be hoped for is that all the energy of motion of the incident particle would be converted into the mass of a new particle.* Thus if the

* Let us assume that the two initial particles are still present at the end of the collision in addition to whatever new particles are produced.

incident and stationary particles both had mass m and the incident particle had energy E, there would be an excess amount of energy $E - mc^2$ over the masses of the two particles initially present, which might be converted into the rest mass of a third particle, which energy conservation would permit to be as big as $(E - mc^2)/c^2$.

However, one can never do as well as this, because momentum must be conserved along with energy. Before the collision one particle is stationary and one is moving, so there is a nonzero momentum equal numerically to $\sqrt{(E/c)^2 - (mc)^2}$ (see (18.38)). After the collision the total momentum must have the same nonzero value, and therefore some of the particles finally present must be moving. This means that some of the excess energy initially available must still be in the form of energy of motion, and consequently not all of it can have been used to create new mass. The pertinent question is how much of the initial energy of motion *can* be converted into new mass?

There are easy and difficult ways of answering this question. The clever method is to answer it in an inertial frame in which the answer is obvious, and then transform the answer back to the frame in which the problem is stated. The answer is obvious in a frame in which the total momentum is initially zero. We can then, as originally suggested, convert *all* the energy of motion into mass of new particles, for this will mean that after the collision all particles are stationary; (i.e., there is no energy left for energy of motion). This is now consistent with conservation of momentum, since the total momentum of a group of particles all with zero velocity is zero. This special frame, in which the total momentum is zero, is known as the center of mass frame, and the frame in which one particle is initially stationary is known as the lab frame.

Since the amount of energy available for conversion into mass of new particles is just the total energy in the center of mass frame minus that part of it ($2mc^2$) already tied up in rest energy, the problem of the particle of maximum mass that can be created by firing a particle of energy E and mass m at a stationary particle of mass m is solved if we can find the total initial energy in the center of mass frame. The answer is then just:

$$\text{Maximum mass} = \frac{E_{CM}}{c^2} - 2m.$$

We can easily compute the total energy in the center of mass frame E_{CM} by exploiting the fact that the square of the total energy minus c^2 times the square of the magnitude of the total momentum is a quantity all observers agree on. (Eq. (18.39)). For an observer in the center of mass frame, the total momentum is zero, so this quantity is just E_{CM}^2. For the lab observer, on the other hand, this quanity is $E_{lab}^2 - c^2 P_{lab}^2$. Now P_{lab} is just the momentum of the moving particle, since the particle at rest has zero momentum. Since for any single particle $E^2 - (pc)^2 = (mc^2)^2$, $P_{lab}^2 = E^2/c^2 - (mc)^2$. Putting these together, we find

$$E_{CM}^2 = E_{lab}^2 - [E^2 - (mc^2)^2]$$
$$= (E + mc^2)^2 - E^2 + (mc^2)^2$$
$$= 2(E + mc^2)mc^2.$$

Hence

$$\frac{E_{CM}}{c^2} - 2m = 2m\left[\sqrt{1 + \frac{1}{2}\left(\frac{E}{mc^2} - 1\right)} - 1\right]. \qquad (18.40)$$

When the velocity of the incident particle is very close to c, as is usually the case in modern accelerators, then the energy of the particle will be much larger than mc^2. Electrons, for example, can be accelerated to energies thousands of times their rest energy. Under these circumstances the 1's appearing in (18.40) can be ignored compared with the very large number E/mc^2, and the formula simplifies to

$$\frac{E_{CM}}{c^2} - 2m = \sqrt{\frac{2mE}{c^2}}, \qquad E/mc^2 \text{ much larger than 1.}$$

This says that as we get to higher and higher energies of the incident particle, the amount of energy that can be converted to rest mass increases only as the square root of the incident energy. Thus, once we have reached the realm of extreme relativistic velocities, in order to increase the amount of mass that can be created by a factor of 10, we have to increase the energy of the incident particle by a factor of 100. Now the more energy available for creating mass, the more we can learn about the elementary particles, and it is therefore an unfortunate consequence of relativity that this energy increases so slowly with incident energy.

In the non-relativistic theory nothing like this happens. The total initial kinetic energy in the lab frame is just $\frac{1}{2}mu^2$. In the center of mass frame the particles have equal and opposite velocities of magnitude $\frac{1}{2}u$, and so the total center of mass kinetic energy is $2[\frac{1}{2}m(u/2)^2] = \frac{1}{2}E_{\text{lab}}$. In the non-relativistic theory this is the energy available for purposes other than motion. It would not occur to the non-relativistic engineer that this might be converted into new mass, but he might use it, for example, to melt one of the particles. The important point, however, is that now the relation between initial energy of motion in the lab and available energy in the center of mass frame is a linear one:

$$E_{CM} = \frac{1}{2}E_{\text{lab}}. \tag{18.41}$$

Hence by increasing the energy of the incident particle one hundredfold, one can increase the amount of energy available for interesting effects by the same factor of 100, as opposed to the disappointing factor of 10 in the relativistic case.

From the point of view of a relativistic engineer, then, if one has a fixed amount of energy to spend and one wants to make as much of it as possible available for the creation of new particles, the most efficient thing to do is to divide the energy in half and fire both particles at each other, for then the lab frame *is* the center of mass frame, and *all* the energy can be converted into mass. By doing this, the available energy improves in the extreme relativistic limit by a factor of $\sqrt{E/2mc^2}$ compared with putting all the energy into one of the particles. In contrast, in the non-relativistic case, this procedure improves things by a factor of only 2. When E/mc^2 is 20,000— about the best one can do today—the improvement in the relativistic case is by a factor of 100.

For this reason people talk of "clashing-beam" accelerators, in which two beams of particles are fired at each other, thus giving more useful energy for the same initial energy investment. So far such machines have not become fashionable, the problem being that it is much harder to make two beams, each with a comparatively small number of particles in it, produce many collisions than it is to make a single beam produce collisions when it is aimed at a large stationary chunk of matter. Nevertheless such machines may be built, and if they are, it will be for purely relativistic reasons.

19
WHY?

If you have read this far, perhaps you now accept and see the consistency in the facts that moving sticks shrink and moving clocks run slowly and fail to be synchronized. It would be surprising, however, if at some points in the argument you did not wonder why these things happen or begin to notice that I have never come to grips with such a question. It is nevertheless an intriguing question and deserves a short, final chapter for its consideration.

Why then, indeed, do moving sticks shrink?*

> * Let us focus our attention on this question. It will be clear that such answers as I can offer apply equally well to the other two questions listed here for your private contemplation:
> Why then, indeed, do moving clocks run slowly?
> Why then, indeed, are moving clocks, spatially separated and synchronized in their proper frame, unsynchronized?

Questions that ask "why" are to be viewed with suspicion, for it is ultimately the business of physics to tell not *why* things happen, but only *what* happens. This has been my attitude throughout this book. It is a fact of nature that moving sticks shrink; our only task is to realize that this happens and to understand why this is a sensible and consistent possibility. Having realized that a universe in which moving sticks shrink (i.e., a universe with finite c) is a possibility just as consistent as one in which moving sticks do not shrink (i.e., with infinite c), we can dispassionately answer the question "Why?" with "Why not?" For the question has been reduced to "Why does c have the value it does?" and there is nothing that assures us that there is a basic reason for the particular numerical value of every important physical constant.*

* Although some physicists hope that ultimately this will turn out to be the case.

This is hardly satisfactory if, notwithstanding the logical necessity for sticks shrinking given the finite value of c, you still find it disturbing that genuine material sticks are so cooperative as to comply with our analysis. If you analyze this uncomfortable sense, you will find that what must lie at the core of it is an implicit theory of what makes sticks as long as they are. The question "Why do moving sticks shrink?" really means "Explain, in terms of the basic notions of stick-length theory, why it is that moving sticks shrink." On that basis I can give an answer, which you may find satisfactory. I shall then immediately end the book, and when, as it will, the question "Why such goings-on in stick-length theory" arises to plague you, I refer you back to the third paragraph of this chapter.

Let us consider then the structure of simplified stick-length theory. Sticks are composed of complicated molecules, which are held together entirely by electromagnetic forces. The length of a stick is determined by the electric (and magnetic) forces between and within molecules. Now it is a rather well known fact that the nature of the electromagnetic forces between moving particles differs from the form these forces have when the particles are at rest. For example, two stationary charged particles exert only electric forces on each other, but when moving, they exert both electric *and* magnetic forces (even if both are moving with the same velocity).

It is a remarkable fact that the electromagnetic forces between the particles of a moving stick differ from those between the particles of a stationary stick in just such a way as to cause the length of the moving stick to be smaller by a factor $\sqrt{1 - (v/c)^2}$. From one point of view this is only to be expected, since if the electromagnetic forces did not have this property, moving sticks would not shrink appropriately, and the principles of relativity and the constancy of the velocity of light would be violated. On the other hand, you may find it reassuring to know that there is a precise and detailed theory*

* Too sophisticated mathematically to be expounded here.

of electromagnetic forces from which this prediction follows.

And that is why moving sticks shrink.

But what about an observer moving with the stick? He finds that it does not shrink. Indeed, if the principle of relativity is not to be violated, he must find that the electromagnetic forces are precisely what we would find they were for a stick at rest. Thus different observers must disagree on what the electromagnetic field (technically a more correct name than force) actually is.

When we add the electromagnetic field to the set of quantities (lengths, times, energies, momenta) that different inertial observers assign different values to, if the principle of relativity is not to be violated, each inertial observer must be given a further set of rules telling him how to calculate from the fields he measures what fields an observer moving past him with speed v will measure. These rules must be such that if A chooses stubbornly to regard his fields as right and B's as wrong, he will nevertheless note that B can regard his own fields as right and yet correctly deduce the fields that A measures by applying the same rules to the fields measured by B.

The detailed way in which this works is an integral part of the very beautiful theory of electromagnetism, which accounts precisely for the shrinking of moving sticks in just the necessary way. Furthermore, insofar as we understand them, the laws of nuclear forces are such that a stick held together by nuclear forces (such as a particular nucleus of a particular size or shape) will shrink according to the Fitzgerald contraction when set into motion. Indeed, we feel so confident that any conceivable force law will lead to the Fitzgerald contraction that in trying to guess what the forces might be like between the new particles being produced in high-energy accelerators, we consider only possibilities meeting this requirement.

In short, the entire body of physical law must enable any observer to support in detail his contention that moving sticks shrink. Therefore in addition to the laws of nature relating various physical quantities, there must always be transformation rules telling us how to relate the quantities measured by one inertial observer to those measured by another moving past the first with speed v. These laws must be symmetric as are Rules 1 through 5, so that given the natural laws and transformation rules of any one observer, we can deduce that any other observer will describe his version of the facts by the same laws and same transformation rules.

This is the structure of physical law as far as it is understood today. I hope some hint of its engrossing and majestic beauty has been suggested here. For it is ultimately just this overwhelming beauty that renders the final, unanswerable "why" superfluous.

PROBLEMS

The first eleven of the problems that follow are based on the first seven chapters. Of the rest, Prob. 12 can be done after Chap. 12; Prob. 13 after Chap. 16; and Probs. 14 through 17 after Chap. 17. The remaining problems all deal with energy and momentum conservation, and are based on Chap. 18.

1. One could, in principle, use the Fitzgerald contraction to measure velocities. Suppose, for example, a train went through a station at a velocity v comparable to c, and two men on the platform, 100 meters apart, synchronized their watches and at time zero touched the point of the speeding train directly in front of them with a pencil. Observers in the train could then measure the distance between the two pencil dots and radio this information back to the station. How far apart will the train observers find the marks to be? By what reasoning will the station observers deduce from this the velocity of the train? Why, given the velocity deduced by the station observers and the fact that they were 100 meters apart in the station, will the train observers agree with their conclusion?

2. A flat, horizontal pan of dough speeds under a circular cookie cutter at a velocity near c. A baker, holding the cutter perfectly horizontally, stamps the dough with lightning speed. (Assume he raises the cutter again so quickly that nothing gets squashed or stuck in the cutter.) The resulting cookie will not be circular, but elliptical. (By the shape of a cookie, one means, of course, its shape in its proper frame.) Is it longer in the direction of its motion or the perpendicular direction? Justify this in the frame of the baker and in the frame of the dough.

3. Prove that a stick of proper length l has a length l' in a frame in which it moves with speed v along a line that makes an angle θ with its length, where

$$l' = l \sqrt{\frac{1 - v^2/c^2}{1 - \sin^2 \theta \ (v^2/c^2)}}.$$

(One way to do this is to consider the stick to be the hypotenuse of a right triangle, one leg of which is parallel and the other perpendicular to the direction of motion.)

4. Two clocks are synchronized by direct comparison in the same place. One is then transported a distance l away to the right at a speed v. By how much must that clock be set ahead in order to be resynchronized? An observer moving with speed u to the left along the line joining the two clocks will say after the experiment that the clock in front is behind the clock in the rear by lu/c^2. On the other hand, since he watched the synchronization experiment and saw the separation, during which the clocks moved apart at speeds u and w, he can deduce directly their asynchronization. Show that he gets the answer lu/c^2 only if $w = (u + v)/(1 + uv/c^2)$. This is therefore yet another way of deriving the addition law.

5. Consider a stick with a mirror on the right end. At a given moment a photon and a particle moving with velocity u leave the left end, moving to the right along the stick. The photon reaches the mirror first, is reflected, and, moving back to the left, encounters the particle still moving to the right at a point whose distance from the left end is a fraction x of the total length of the stick. Using the fact that x is an invariant quantity, show using only the constancy of the velocity of light and the principle of relativity (i.e., it is not necessary to assume anything about the shrinking factor for the moving stick) that an observer moving to the left with velocity v with respect to the stick will find the velocity of the particle to be

$$w = \frac{u + v}{1 + uv/c^2} .$$

6. If a train of length l is stationary, then light from a bulb flashed at the midpoint of the train will reach the two ends at the same time. If the train moves past a station platform at speed v, then (in the frame of the station) the light from a single flash of this bulb will no longer reach the two ends of the train at the same time, since the rear end is moving toward the light while the front end is moving away. Evidently if we want to place the bulb so the light reaches the ends of the train simultaneously in the frame of the station, we must move the bulb closer to the front of the train; i.e., we must attach it at a point a fraction x of the total length of the train behind the front, where x is less than $\frac{1}{2}$. What is x? (Nothing is required to do this but the fact that the speed of light is c in the frame of the station.) The answer is $x = (\frac{1}{2}) (1 - v/c)$, which reduces to the obvious answer $x = \frac{1}{2}$ when $v = 0$ and has the very reasonable property of requiring the bulb to get closer to the front of the train, as v gets closer to c.

7. Although nothing was used but the constancy of the velocity of light in deriving the result of Prob. 7, from this result we can immediately deduce Rule 4. For consider two clocks synchronized in the train frame, one at each end of the train. According to station observers the light from the bulb reached each clock at the same time. However, according to observers on the train, the light from the bulb obviously reached the front clock before it reached the one in the rear, since the bulb is attached to the train at a point closer to the front. (The speed of light is c in the train frame too.) Show, using the result of Prob. 7, that in the train frame the light reaches the rear clock a time lv/c^2 after it reaches the clock in front, where l is the length of the train in its proper frame. Why does Rule 4 follow?

8. Deduce the Fitzgerald contraction using only Rule 4 and the principle of relativity as follows: Consider a train of proper length l stationary in the station and another train moving through the station with speed v and proper length l/a_v, where a_v is the shrinking factor which we wish to prove equals $\sqrt{1 - (v/c)^2}$. The proper length of the moving train has been chosen so that to observers in the station it appears to have a length $a_v(l/a_v) = l$, that is, precisely the length of the stationary train. Thus when the clock on each end of the stationary train reads zero, an end of the moving train is over each. Noting the readings of the clock on each end of the moving train at the moment each is over the appropriate end of the stationary train, show that an observer on the moving train will conclude that the length of the stationary train is (l/a_v) $[1 - (v/c)^2]$. Since the principle of relativity requires him to have a rule that moving trains shrink by a factor a_v and since the proper length of the train in the station is l, this quantity must be $a_v l$, from which $a_v = \sqrt{1 - (v/c)^2}$ follows.

9. In a manner similar to Prob. 9, deduce the slowing down of moving clocks using only Rule 4 and the principle of relativity.

10. Deduce from Fig. 6.2, page 52, the velocity of the stick. All necessary data are in the figure itself.

11. A 2-meter stick slides horizontally on ice so fast that it contracts to $\frac{1}{2}$ meter and falls through a hole in the ice 1 meter in diameter. Describe this catastrophe in the frame of the stick.

12. Show that the hypothetical rules given in the footnote on page 113, Chap. 12 are all consistent with the principle of relativity. Can you think of a simple reason why they could not possibly be found in nature?

13. Prove that at the moment of B's transfer (Chap. 16), inertial observers in the neighborhood of Vega could have been found maintaining that the time on Earth was anything between $1\frac{1}{2}$ months and 50 years $1\frac{1}{2}$ months after departure.

14. Consider the crash of the faster-than-light ship of Chap. 15. In a Minkowski diagram in which the Earth-Sirius frame uses the rectangular coordinates, draw the world line of the ship from a long while before takeoff on Earth to a long while after the crash near Sirius. Then draw in the lines of constant t' for an observer (moving slower than light) who describes things as in the footnote on page 137.

15. Draw Minkowski diagrams describing Prob. 8 and your solution to Prob. 9. Deduce the addition law for parallel velocities from a Minkowski diagram containing, in addition to the perpendicular axes, axes used by observers moving to the left and right. (It is not necessary to use the scaling factor.)

16. Solve Prob. 13 using a Minkowski diagram.

17. Problem 11 is readily solved by constructing a 3-dimensional Minkowski diagram in which the two spatial dimensions describe the plane perpendicular to the ice containing the stick. (The third dimension is, of course, the temporal one.) Sketch (or construct with cardboard, scissors, and glue) the diagram.

18. A rocket of mass m initially at rest carries a quantity of fuel of mass M. It converts an amount M_1 of this fuel into energy which it uses to throw the remaining mass of fuel M_2 ($M = M_1 + M_2$) backward. What energy of motion does the rocket acquire in the process? Answer: $E = (M^2 - M_2^2)c^2/2(M + m)$. Note that the most efficient thing to do is therefore to convert as much as possible of the fuel into energy. Ideally all the rest mass should be converted into energy and then zero-mass particles (such as photons) hurled backward with this amount of energy.

19. Show that to reach a velocity close to c, the mass of the fuel must be much greater than the mass of the ship in Prob. 18.

20. Do the exercise suggested in the footnote on page 210.

21. Verify that $E^2 - c^2 (P_x^2 + P_y^2 + P_z^2)$ is independent of which inertial observer calculates E and the components of P, as asserted in (18.39), page 221.

22. Generalize (18.40), page 223, to the case in which the original two particles do not have the same mass. (If the particle at rest in the lab has mass m, and the incident particle mass m', the solution is:

$$\frac{E_{CM}}{c^2} - (m + m') = (m + m')$$

$$\left(\sqrt{1 + \frac{2mm'}{(m + m')^2} \left(\frac{E}{m'c^2} - 1 \right)} - 1 \right).$$

23. Derive the non-relativistic limit (18.41) from (18.40). What is the non-relativistic limit of the answer to Prob. 22? Can you derive it directly by non-relativistic arguments?

24. What is the form of the answer to Prob. 22 when m is much larger than m' or E/c^2, and why is this reasonable?

INDEX

INDEX

Absolutist view, consistency of, 84, 85, 91–97
Accelerators, 9, 77, 201, 224, 228
Addition law for velocities, non-relativisitic, 129, 131, 204
  relativistic, for non-parallel velocities, 206
  for parallel velocities, 129–134, 206, 230, 231, 233
Aristotle, 1–3

c, 10
Causality, principle of, 138
Center of mass frame, 222
Chemical energy, 220
Chemical explosives, 220
Clock, light beam (see Light beam clock)
Clock paradox, 40, 41, 141–153, 187–194
Clocks, non-relativistic, imitating relativistic effects, 99–117
  pre-relativistic assumptions about, 21
  slow rate of when moving, 33–46
  consistency of, 87–90, 186, 187
  observation of, 42
  reason for, 225–228
  (See also Rule Two)
Coincidences, principle of invariance of, 24, 25, 50, 51, 81, 128

Collisions, 202
  inelastic, 217
Conservation of energy and momentum, law of, 201–204
  (See also Energy; Energy and momentum; Momentum)
Conservation of mass, 218
Constancy of velocity, of blips, principle of, 16, 17
  of light, principle of, 1, 2, 9–17, 19, 20, 23, 39, 114–116
  of neutrinos, principle of, 16
Coordinate system, 120, 121, 160–164
  oblique, 162
  rectangular, 161, 162
Cosmic rays, 9, 42, 77

$E = mc^2$, 201, 215
Einstein, A., 2, 9, 10, 13, 119
Electromagnetic forces, 55, 202, 220, 226, 227
Electromagnetism, connection with relativity, 9, 119
Electrons, 9, 77
Elementary particles, 9, 77, 223, 228
Energy, chemical, 220
  electromagnetic, 220
  of motion, 218
  non-relativistic, 203
  nuclear, 220
  relativistic, 212

Energy, thermal, 216, 218
  total, 204
Energy and momentum, 201–224,
    233, 234
  law of conservation, 201–224
    applications of, 221–224, 233,
      234
    difference between relativistic
      and non-relativistic law,
      215–219
    non-relativistic law, 202, 205
    non-relativistic form, 203, 204,
      211
    relativistic form, 212
    transformation rules, non-rela-
      tivistic, 205
    relativistic, 213
Engineer, relativistic, 201, 221,
    224
Estheticism, 228
Ether, 12, 13, 134
Experiments bearing on relativity,
    2, 3, 13, 42, 133, 134, 201,
    221–224

Fitzgerald, G. F., 55
Fitzgerald contraction, 47–62, 174,
    229, 232
  consistency of, 79–86, 90, 174,
    183–186
  reason for, 225–228
  (See also Rule Three)
Four-velocity, 208, 209
Fourth dimension, 199

Galileo, and principle of relativity,
    1–3
  and velocity of light, 10
General relativity, 142, 151–153
Gravitation, 152, 153, 202

Index of refraction, 133
Inelastic collisions, 217
Inertial, 6, 7
Invariant, 207

Kinetic energy, 203

Lab frame, 222
Language, difficulties and ambigui-
    ties in, 22, 23, 27, 28, 33, 40,
    41, 56, 77, 78, 86, 90, 137,
    141, 148, 172, 185, 186, 196
Length of a stick, moving at an
    angle to itself, 47, 230
  moving parallel to itself, 47–62,
    79–86
  moving perpendicular to itself,
    27–32
  (See also Fitzgerald contraction;
    Meter sticks)
Light, peculiar role in relativity of,
    9, 16
  (See also Constancy of velocity,
    of light; Velocity of light)
Light beam clock, moving perpen-
    dicular to itself, 33–38
  used to measure length of stick,
    49–54
Lorentz, H. A., 55, 119
Lorentz contraction, 55
  (See also Fitzgerald contraction)
Lorentz transformation, 119–128
  applications of, 136, 155, 205,
    206, 210
  best units for, 180
  derived from Minkowski dia-
    gram, 175–180
  formulas, 126
  utility of, 120, 127, 128

Mass, 203, 209
  deficit, 220
  law of conservation, 218
  relativistic, 220
Meter sticks, non-relativistic, imi-
    tating relativistic effects, 99–
    117
  non-relativistic assumptions
    about, 21

Meter sticks, as region of space-time, 171, 172
(*See also* Fitzgerald contraction; Length of a stick)
Michelson-Morley experiment, 13
Minkowski, H., 155
Minkowski diagrams, 155–199, 232, 233
applications of, 175–199, 232, 233
to asynchronization of moving clocks, 180–183
to clock paradox, 187–194
to Fitzgerald contraction, 183–186
to Lorentz transformation derivation, 175–180
to pole-in-barn paradox, 194, 195
to slowing down of moving clocks, 186, 187
scale of axes, 169–174
scaling hyperbola, 174, 175
space axis, orientation of, 168
time axis, orientation of, 165
three- or four-dimensional, 198, 199, 233
Molecules, 220
Momentum, non-relativistic, 204
relativistic, 209, 210, 212
total, 204
(*See also* Energy and momentum)
Mössbauer effect, 42
Mu mesons, 42

Natural units, 180
Neutrinos, 16
Neutrons, 220
Newton, I., 7
Newton's first law of motion, 7, 202
Non-relativistic model of relativistic effects, 99–117, 138–140
Non-uniform motion, 4, 5, 7, 142, 143, 152

Nuclear forces, 202, 220, 228
Nuclear weights, 220
Nuclei, 42, 219

Paradox, clock, 141–153, 187–194
how to avoid, 90
(*See also* Language)
pole-in-barn, 194, 195
shrinking sticks, 56, 79–86
slowly running clocks, 40, 41, 87–90
twin, 141–153, 187–194
Particles, elementary, 9, 77, 223, 228
Photons, 15
Potential energy, 203
Principle, of causality, 138
of constancy of velocity of light (*see* Constancy of velocity, of light)
of invariance of coincidences (*see* Coincidences)
of relativity (*see* Relativity)
of rotational invariance, 31, 44, 50, 67
Proper, 56, 57
Proper time, 208
Protons, 9, 77, 220

Radioactivity, 219
Refraction, index of, 133
Relativistic effects, difficulty of observing, 75–77
importance of, 77
Relativistic mass, 220
Relativity, general theory of, 142, 151–153
principle of, 1–7, 9, 20, 32, 38, 125, 170
of simultaneity, 66, 67, 79, 90, 148
(*See also* Rule Four)
why the name, 5

Rotational invariance, principle of, 31, 44, 50, 67
Rule One, 32, 73, 126
Rule Two, 38, 39, 73, 186, 187
Rule Three, 55, 73, 183–186
Rule Four, 66, 67, 74, 137, 148, 180–183, 231
Rule Five, 68, 74, 126

Simultaneity, relativity of, 66, 67, 79, 90, 148
    (*See also* Rule Four)
Space (*see* Fitzgerald contraction; Length of a stick; Meter sticks)
Speed of light (*see* Velocity of light)
Synchronization of moving clocks, 63–72
    how to remember which is behind, 72
    (*See also* Rule Four; Rule Five)

Teleology, 225–228
Thermal energy, 216, 218

Time (*see* Clocks)
Time dilation (*see* Clocks; Rule Two)
Time ordering, 137, 138
Total energy, 204, 213
Total momentum, 204, 213
Trajectory, space-time, 157
Twin paradox (*see* Clock paradox)

Uniform motion, 4–6

Velocities faster than light, 12, 37, 132, 133, 135–140, 212
Velocity of light, constancy of (*see* Constancy)
    independent of source, 14
    numerical value of, 10
    significance of, in relativity, 9, 16, 114–116
    in water, 133

World line, 157
    of object with given velocity, 159